Nenad Marjanovic

Photoresponsive Organic Field-Effect Transistors
(photOFETs)

Nenad Marjanovic

Photoresponsive Organic Field-Effect Transistors (photOFETs)

Photodoping in OFETs

VDM Verlag Dr. Müller

Imprint

Bibliographic information by the German National Library: The German National Library lists this publication at the German National Bibliography; detailed bibliographic information is available on the Internet at http://dnb.d-nb.de.

Cover image: www.purestockx.com

Publisher:
VDM Verlag Dr. Müller Aktiengesellschaft & Co. KG , Dudweiler Landstr. 125 a, 66123 Saarbrücken, Germany,
Phone +49 681 9100-698, Fax +49 681 9100-988,
Email: info@vdm-verlag.de

Zugl.: Linz, Johannes Kepler University Linz, Austria, Diss., 2006

Produced in USA and UK by:
Lightning Source Inc., La Vergne, Tennessee, USA
Lightning Source UK Ltd., Milton Keynes, UK
BookSurge LLC, 5341 Dorchester Road, Suite 16, North Charleston, SC 29418, USA

ISBN: 978-3-8364-8367-4

Preface

The presented work was carried out at the Linz Institute for Organic Solar Cells (LIOS), Physical Chemistry, Johannes Kepler University Linz, between January 2003 and November 2005.

I want to thank Professor Niyazi Serdar Sariciftci for supervising this work and for guiding me through this period. Also, I want to thank Assistant Professor Helmut Neugebauer for fruitful discussions and help.

I thank Dr. Birendra Th. Singh, for his very clear advice and great friendly support.

I thank especially my colleagues Dr. Gilles Dennler, fellow PhD students Serap Günes and Robert Koeppe for their participation in valuable experiments and discussions. Also, I thank Dr. A. Mozer and Dr. G. Matt for fruitful discussions.

I thank all former and present members of our Institute: Dr. Dieter Meissner, Dr. Farideh Meghdadi, DI Christoph Lungenschmied, DI Anita Fuchsbauer, Dr. Le Huong Nguyen, DI Martin Egginger, Hans-Jürgen Prall, Dr. Shengli Lu, DI Sandra Hofer, DI Daniela Stoenescu, Dr. Andrei Andreev, Dr. Elif Arici, Dr. Harald Hoppe, Dr. Christoph Winder, and Dr. Martin Drees, for many fruitful discussion and suggestions.

I thank our collaborators Professor Siegfried Bauer and Dr. Reinhard Schwödiauer (Soft Matter Physics Department, JKU), for many fruitful discussions, suggestions, and support.

I want to thank especially my family and my parents for having understanding and patience with me during this period.

DI Dr. Nenad Marjanović
Linz, February 2008.

Contents

Chapter 1

1. Introduction

Background and Motivation

Photodetectors are semiconductor-based devices that can convert optical signals into electrical signals. The operation of a photodetector involves three steps: charge carrier generation by absorption of the incident light, charge carrier transport, and charge carrier collection by electrodes.

Photodetectors include *photoconductors*, *photodiodes/photovoltaic devices (e.g. solar cell)* and, to some extent, *phototransistors*. Photodetectors have a broad range of applications including, *e.g.* sensors or detectors, power converters or image sensors.

A *photoconductor* consists of a semiconducting material sandwiched between two ohmic contacts. When incident light impinge on the photoconductor, electron-hole pairs are generated. This corresponding increase in the number of charge carriers results in an increase of the conductivity. The photocurrent flowing between the contacts depends on the electric field inside the photoconductor and on the carrier drift velocity.

A *photodiode* is basically a p-n junction or a metal-semiconductor contact operating under reverse bias. When an optical signal impinges on the photodiode, the electric field present in the depletion region separates the photogenerated electron-hole pairs and an electric current flows in the external circuit.

A *solar cell* is similar to a photodiode and follows the same operating principle. However, the solar cell is operated in the forward bias direction (Maximum Power Point, mpp). Usually it is a large-area device and in addition able to absorb the largest possible part of the spectrum. The photovoltaic effect developed under illumination results in a power conversion of the solar electricity (power delivered to the load per incident solar energy), which can be extracted from the device.

Phototransistors combine the two above-mentioned photoinduced effects (*i.e.* photoconductivity and photovoltaic effect) with transistor action. Therefore, a phototransistor can have high gains. The output photocurrent depends on the gate voltage and on the illumination intensity.

Thin film phototransistors based on inorganic semiconductors [1-2] or various organic and polymeric semiconductors, such as poly (3-octylthiophene), polyfluorene, bifunctional spiro compounds, polyphenyleneethynylene derivative or 2,5-bis-biphenyl-4-yl-thienol[3,2-b]thiophene (BPTT) [3-7] were reported. However, phototransistors based on conjugated polymer/fullerene blends, have not been demonstrated until now.

The main topic of this work is the realisation and characterisation of photoresponsive Organic Field-Effect Transistors (photOFETs) based on conjugated polymer/fullerene blends for the photoactive semiconductor layer, and highly transparent organic polymeric gate-dielectrics.

Organic Thin Film Transistors

1.2.1. Operating principles of Organic Thin Film Transistors

Weimer introduced the concept of the Thin Film Transistor (TFT) in 1962 [8]. Since then, this device concept has been adapted to low conductivity materials, and is now commonly used in amorphous silicon technology [9-10].

In principle, the TFT is an insulating gate device; it operates in the accumulation regime, rather than in the inversion regime typical for crystalline-Si technology [11].

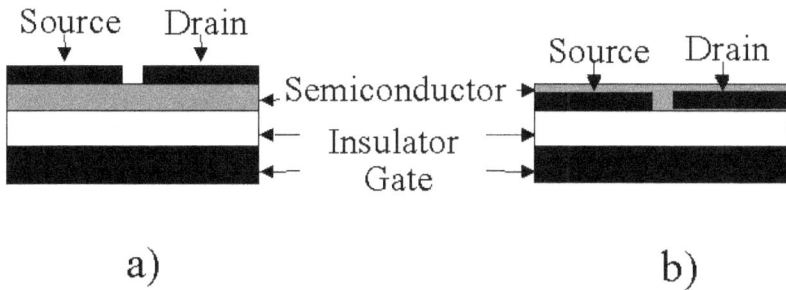

Fig. 1.1. Schematic of top **a)** and bottom **b)** contact Organic Thin Film Transistors.

As shown in Fig. 1.1, there are two basic schemes for organic thin film transistors. In both arrangements an organic semiconductor film is deposited on a gate-insulator layer and is contacted with metallic source and drain electrodes. Ideally the source and drain should form an ohmic contact with the active semiconductor.

The geometrical device parameters are the source - drain channel length (L), the channel width (W), and the insulator capacitance per unit area, C_{ins}, Fig. 1.2.

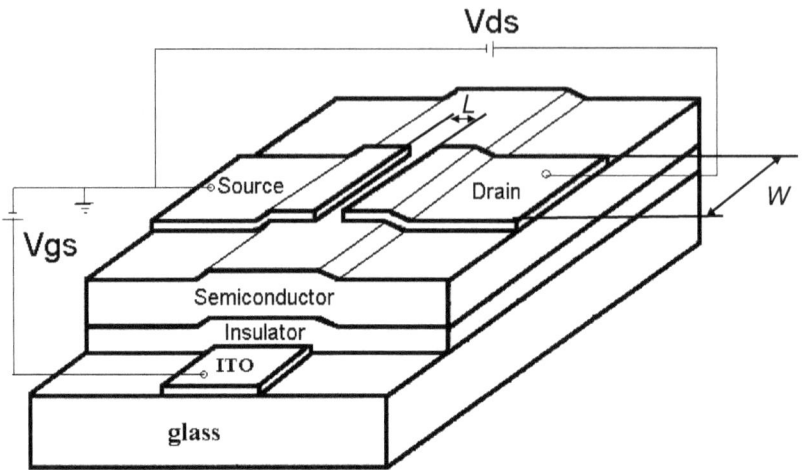

Fig. 1.2. Schematics of the TFT connection.

The voltage applied between the source and drain contacts is referred to as the source-drain voltage, V_{ds}. Generally, for a given V_{ds}, the amount of current that flows through the semiconductor film from the source to the drain contact is a function of the gate-source voltage, V_{gs}. In a phototransistor, the drain-source current may also depend on the illumination. The semiconductor film and the gate electrode are capacitively coupled such that an application of a bias voltage on the gate induces a charge modulation at the insulator/semiconductor interface. Most of these charges are mobile and move according to the applied source-drain voltage, V_{ds}. Ideally, when no gate voltage is applied, the conductance of the semiconductor film is low because there are no mobile charge carriers; *i.e.*, the device is in the "off-state". When a suitable gate voltage is applied, mobile charges are accumulated, and the transistor is in the "on-state". The source contact is connected to ground.

Two different methods are commonly employed for the characterization of TFTs: Either V_{gs} is kept constant and V_{ds} is swept (output curves, Fig.1.3.a) or V_{ds} is held constant and V_{gs} is swept (transfer characteristics, Fig.1.3.b).

8

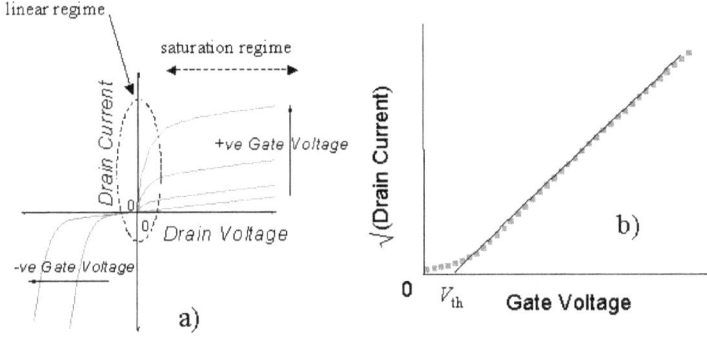

Fig. 1.3. a) Output curves of TFT working in electron-enhanced mode;
b) Transfer curve for the transistor operating in electron enhanced mode plotted as $\sqrt{I_{ds}}$ vs. V_{gs}.

As can be seen in Fig.1.3, in the case of positive applied voltages V_{ds} and V_{gs}, an electron-enhanced mode is developed. For hole-enhanced mode operation, the bias voltages are negative. The intercept of the extrapolated linear curve with the gate voltage-axis in the transfer characteristics shown in Fig. 1.3(b) defines the threshold voltage, V_{th}. The drain-source current in the linear and in the saturation regimes are given by Equations (1) and (2), respectively [11]:

$$I_{ds} = \frac{W}{L} C_{ins} \mu \left[\left(V_{gs} - V_{th} \right) V_{ds} - \frac{V_{ds}^2}{2} \right],$$ (1)

$$I_{ds} = \frac{W}{2L} C_{ins} \mu \left(V_{gs} - V_{th} \right)^2,$$ (2)

where μ is field-effect mobility. Equation (1) describes the transport in the case when $V_{ds} <$ $(V_{gs} - V_{th})$. When $V_{ds} > (V_{gs} - V_{th})$, equation (2) is valid. The field-effect mobility, μ, can be calculated either from the linear or from the saturation regime. The mobility, μ is largely determined by the morphology of the semiconductor film at the insulator/semiconductor interface [12]. In addition to μ and V_{th}, another important device characteristic is the on-off drain-source current ratio, I_{on}/I_{off}, which is basically dependent on the device geometry.

The threshold voltage in the accumulation regime is given by Equation (3) [13]:

$$V_{th} = \pm \frac{q n_o d}{C_{ins}} + V_{FB},$$ (3)

where V_{FB} is the flat-band potential, q is elementary charge, n_0 is the bulk carrier density, and d is the thickness of the semiconductor. The sign of the right-hand side in equation (3) corresponds to the sign of the majority carriers.

1.2.2. Operating principles of photoresponsive Organic Field-Effect Transistors (photOFETs)

Upon illumination two different effects are observed in the active layer of transistors, *i.e. photoconductivity* and the *photovoltaic effect*. When the transistor is in the ON-state the photocurrent is dominated by the photovoltaic effect and is given by Equation (4) [14]:

$$I_{ph,pv} = G_M \Delta V_{th} = \frac{AkT}{q} \ln\left(1 + \frac{\eta q \lambda P_{opt}}{I_{pd} hc}\right) \qquad (4)$$

where η is the quantum efficiency (i.e. the number of charge carriers generated per incident photon), q is the elementary charge, P_{opt} the incident optical power, I_{pd} the dark current for electrons, hc/λ the photon energy, G_M the transconductance, ΔV_{th} the threshold voltage shift, and A is a fit parameter. The photovoltaic effect together with the shift of the threshold voltage is caused by the large number of trapped charge carriers under the source contact [3-7]

When the transistor is in the OFF-state, the photocurrent is dominated by photoconductivity as described by Equation (5) [15]:

$$I_{ph,pc} = (q\mu n E)Wd = BP_{opt}, \qquad (5)$$

where μ is the charge carrier mobility, n is the carrier density, E the electrical field in the channel, W the gate width, and d the thickness of the active layer. $I_{ph,pc}$ is therefore directly proportional to P_{opt} with a proportionality factor B.

The photocurrent is characterized by a high gain and fast saturation especially at low illumination intensities.

Useful figures-of-merit for phototransistor are:

- The *responsivity*, R (expressed in A/W) of the device, which is defined as [4]:

$$R = \frac{I_{ph}}{P_{opt}} = \frac{(I_{dsillum} - I_{dsdark})A^{-1}}{P_{inc}}, \qquad (6)$$

where I_{ph} is the drain-source photocurrent, P_{opt} is the incident optical power, P_{inc} the power density of the incident light per unit area, $I_{ds,illum}$ the drain-source current under illumination, $I_{ds,dark}$ the drain-source current in the dark and A the effective device area.

- The *photosensitivity*, P or signal (photocurrent) to background (dark current) ratio of the device, which is defined as [4]:

$$ P = \frac{signal}{background} = \frac{I_{ph}}{I_{dsdark}} = \frac{\left(I_{dsillum} - I_{dsdark} \right)}{I_{dsdark}}. \tag{7} $$

- The *photoresponse*, $R_{I/d}$, or the ratio of the total drain-source current under illumination to the drain-source current in the dark, which is defined as [4]:

$$ R_{I/d} = \frac{I_{dsillum}}{I_{dsdark}}. \tag{8} $$

The photoresponse exhibits a power law dependence on the illumination according to Equation [4]:

$$ R_{I/d} \propto P_{inc}^{\alpha}. \tag{9} $$

In equation (9), P_{inc} is the power density of the incident light per unit area and the α is the exponent, which is a function of the applied voltage V_{gs} [4].

1.3 Conjugated polymer/fullerene bulk heterojunctions

The observation of a photoinduced electron transfer from optically excited conjugated polymers to C_{60} molecule and the observation of highly increased photoconductivities upon C_{60} addition to conjugated polymers led to the concept of a polymer/fullerene bulk heterojunction [16-24]. The photoinduced electron transfer occurs when it is energetically favourable for the electron in the S_1-excited state of the polymer to be transferred to the much more electronegative C_{60}, thus resulting in an effective quenching of the excitonic photoluminescence of the polymer [16]. The photoinduced charge transfer is depicted schematically in Fig. 1.5, together with an energy level representation [25].

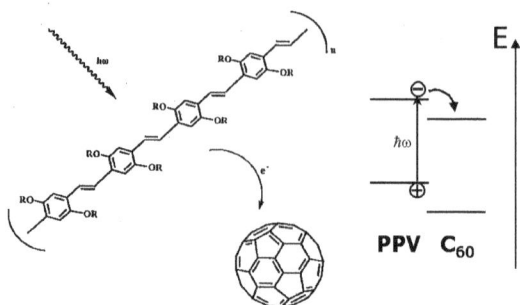

Fig. 1.5. Illustration of the photoinduced charge transfer (left) with a sketch of the energy level scheme (right).

The bulk heterojunction concept was introduced by blending two organic semiconductors having electron donor (D) and electron acceptor (A) properties in solution [26-28]. Spin cast films from such binary solutions then resulted in solid-state mixtures of organic semiconductors.

The essence of the bulk heterojunction is to intimately mix the donor and acceptor components in the volume bulk so that each donor-acceptor interface is within a distance less than the exciton diffusion length of each absorbing site. In Fig.1.5, right, the situation is schematically shown for a bulk heterojunction device, again neglecting all kinds of energy level alignments and interface effects. The bulk heterojunction is similar to a bilayer device with respect to the D-A concept, but it exhibits a largely increased interfacial area where charge separation is occuring. Due to the interface being dispersed throughout the bulk, no loss due to a too small exciton diffusion length is expected, because ideally all excitons will be dissociated within their lifetime. In this concept the charges might also be separated within the different phases and hence recombination is reduced to a large extent and the photocurrent often follows the light intensity linearly or slightly sublinearly [25]. The bulk heterojunction requires percolated pathways for the hole and electron transporting phases to the contacts. In other words, the donor and acceptor phases have to form a bicontinuous and interpenetrating network. Therefore, bulk heterojunction devices are much more sensitive to the nanoscale morphology in the blend.

The organic solar cell power conversion efficiency in devices based on bulk heterojunction conjugated polymer/fullerene blends (P3HT: PCBM) reaches ~ 5% under AM1.5 (80 mW/cm^2) [29, 30].

1.4 References

[1] Y. Kaneko, N. Koike, K. Tsutsui, and T. Tsukada, *Appl. Phys. Lett.* 56, 650, 1990.

[2] H.-S. Kang, C.-S. Choi, W.-Y. Choi, D.-H. Kim and K.-S. Seo, *Appl. Phys. Lett.* 84, 3780, 2004.

[3] K. S. Narayan and N. Kumar, *Appl. Phys. Lett,* 79, 1891, 2001.

[4] M. C. Hamilton, S. Martin, and J. Kanicki, *IEEE Trans. Electron Devices*, 51, 877, 2004.

[5] T. P. I Saragi, R. Pudzich, T. Fuhrmann, and J. Salbeck, *Appl. Phys. Lett.* 84, 2334, 2004.

[6] Y. Xu, P. R. Berger, J. N. Wilson, and U. H. F. Bunz, *Appl. Phys. Lett.* 85, 4219, 2004.

[7] Y.-Y Noh, D.-Y Kim, Y. Yoshida, K. Yase, B.-J. Jung, E. Lim, and H.-K. Shim, *Appl. Phys. Lett.* 86, 043501, 2005.

[8] P.K. Weimer, *Proc. IRE*, 50, 1462, 1962.

[9] M.J. Powell, B.C. Easton, and O.F. Hill, *Appl. Phys. Lett.* 38, 794, 1981.

[10] T.L. Credelle, *Proceedings of the International Display Research Conference*, San Diego, (IEEE, New York), p. 208, 1988.

[11] S. M. Sze, *Physics of Semiconductor Devices*, Wiley, New York, 1981.

[12] F. Dinelli, M. Murgia, P. Levy, M. Cavallini, F. Biscarini, and D.M. de Leeuw, Phys. Rev. Lett. 92, 116802, 2004.

[13] G. Horowitz, in *Semiconducting Polymer: Chemistry, Physic and Engineering*, edited by G. Hadziioannou, and P.F. van Hutter (Wiley-VCH, Weinheim), 1999.

[14] H-S. Kang, C.S. Choi, W.-Y. Choi, D.-H. Kim, and K.-W. Seo, *Appl. Phys. Lett.* 84, 3780, 2004.

[15] S. M. Sze, *Physics of Semiconductor Devices*, Wiley, New York, p. 744, 1981

[16] N.S. Sariciftci, L. Smilowitz, A.J. Heeger, and F. Wudl, *Science* 258, 1474, 1992.

[17] L. Smilowitz, N.S. Sariciftci, R. Wu, C. Gettinger, A.J. Heeger, and F. Wudl, *Phys. Rev. B* 47, 13835, 1993.

[18] C.H. Lee, G. Yu, D. Moses, K. Pakbaz, C. Zhang, N.S. Sariciftci, A.J. Heeger, and F. Wudl, *Phys. Rev. B* 48, 15425, 1993.

[19] S. Morita, A.A. Zakhidov and K. Yoshino, *Solid State Commun.* 82, 249, 1992.

[20] S. Morita, S. Kiyomatsu, X.H. Yin, A.A. Zakhidov, T. Noguchi, T. Ohnishi, and K. Yoshino, *J. Appl. Phys.* 74, 2860, 1993.

[21] N.S. Sariciftci, D. Braun, C. Zhang, V.I. Srdanov, A.J. Heeger, G. Stucky, and F. Wudl, *Appl. Phys. Lett.* 62, 585, 1993.

[22] L.S. Roman, W. Mammo, L.A.A. Petterson, M.R. Andersson, and O. Inganäs, *Adv. Mater.* 10, 774, 1998.

[23] G. Yu, J. Gao, J.C. Hummelen, F. Wudl and A.J. Heeger, *Science* 270, 1789, 1995.

[24] 50. C.Y. Yang, and A.J. Heeger, *Synth. Met.* 83, 85, 1996.

[25] H. Hoppe and N.S. Sariciftci, *J. Mater. Res.* 19, 1924, 2004.

[26] G. Yu and A.J. Heeger, *J. Appl. Phys.* 78, 4510, 1995.

[27] J.J.M. Halls, C.A. Walsh, N.C. Greenham, E.A. Marseglia, R.H. Friend, S.C. Moratti, and A.B. Holmes, *Nature,* 376, 498, 1995.

[28] K. Tada, K. Hosada, M. Hirohata, R. Hidayat, T. Kawai, M. Onoda, M. Teraguchi, T. Masuda, A.A. Zakhidov, and K. Yoshino, *Synth. Met.* 85, 1305, 1997.

[29] W. Ma, C. Yang, X. Gong, K. Lee, and A.J. Heeger, *Adv. Funct. Mater.* 15, 1617, 2005.

[30] J.Y. Kim, S.H. Kim, H.-H. Lee, K. Lee, W. Ma, X. Gong, and A.J. Heeger, *Adv. Mater.* in press

Chapter 2

2. Experimental

Materials

2.1.1. Conjugated polymer/fullerene blends

The material combination of poly-para-phenylene vinylene and fullerene is probably one of the best investigated blends of conjugated polymer and a fullerene. In the past, several groups investigated bulk heterojunctions of MDMO-PPV and PCBM by (1:4 wt./wt.) either studying the electrical properties of the blend or by using the blend as the active layer in organic solar cells or in organic field effect transistors [1-16].

The MDMO-PPV: PCBM (1:4) blend is also employed in this thesis as a photoactive organic semiconductor.

2.1.1.1. MDMO-PPV and PCBM

Figure 2.1, shows two of the most frequently used and best known organic semiconductors for organic electronic applications, namely poly-[2-methoxy-5-(3',7'-dimethyloctyloxy)]-para-phenylene vinylene (MDMO-PPV) [17] and the soluble derivative of $C_{60,}$ 1-(3-methoxycarbonyl)propyl-1-phenyl-[6,6]methanofullerene (PCBM) [18].

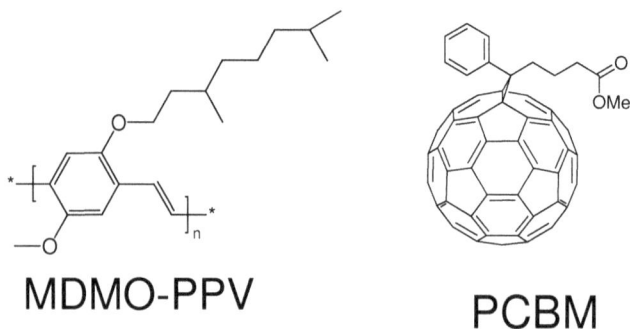

MDMO-PPV   **PCBM**

Fig. 2.1 Structure of poly-[2-methoxy-5-(3',7'-dimethyloctyloxy)]-para-phenylene vinylene (MDMO-PPV) and 1-(3-methoxycarbonyl)propyl-1-phenyl-[6,6]methanofullerene (PCBM).

The absorption spectra of the active MDMO-PPV: PCBM (1:4 wt./wt.) layer shows a strong mismatch to the solar photon flux, as depicted in Fig. 2.2 [19].

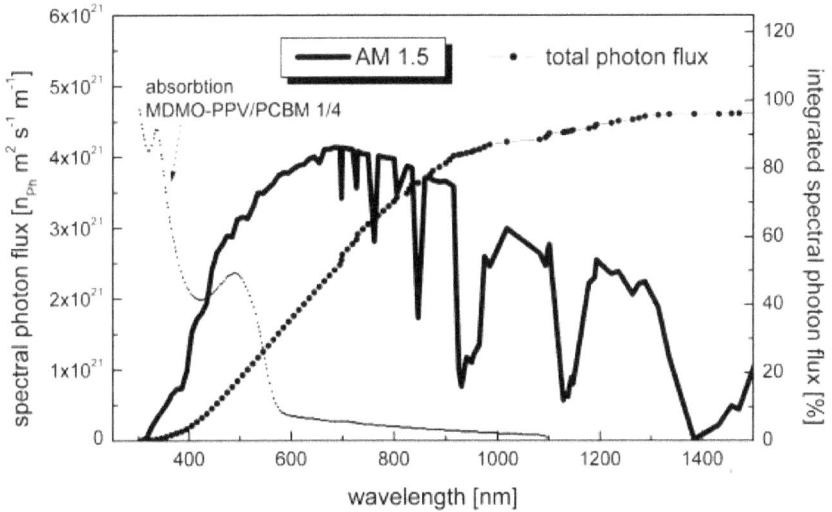

Fig. 2.2 The terrestrial AM1.5 sun spectrum (—) and the integrated spectral photon flux (starting from 0 nm) (-•-) in the comparison with the absorption spectrum (---) of a MDMO-PPV: PCBM (1:4) blend.

In general, two important parameters of a material are the band gap and the band energy levels. The knowledge of these parameters is necessary for engineering devices. The optical band gap E_g^{opt} can be determined from the absorption or luminescence onset.

Fig. 2.3 shows the absorption coefficient of the three different components: MDMO-PPV, PCBM, and blend of MDMO-PPV: PCBM (1:4) [20]. The absorption coefficient of the pristine **MDMO-PPV** film (dotted curve) is shown in Fig. 2.3. The onset is around 580 nm, corresponding to 2.1 eV.

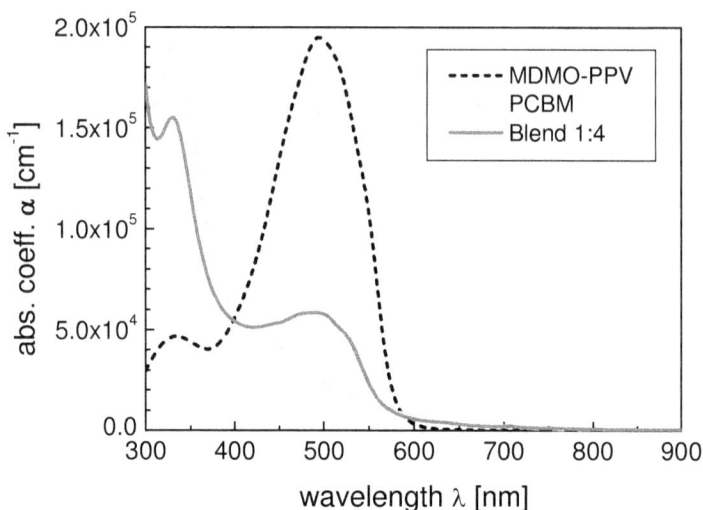

Fig. 2.3 Absorption coefficient of MDMO-PPV (dotted), PCBM (dash), and MDMO-PPV: PCBM (1:4) blend (solid line), calculated from their respective dielectric functions.

Electrochemical measurements can be also used as an alternative method for the determination of the band gap of a material.

The electrochemical band gap E_g^{EC} is defined as the difference of the oxidation and reduction potential. Table 1.1 shows the values for MDMO-PPV and PCBM [19].

Table 1.1 Energy levels of MDMO-PPV and PCBM, calculated from electrochemical measurements

	HOMO / eV	LUMO / eV	E_g^{EC} / eV	E_g^{Opt} / eV
MDMO-PPV	5.3	3.0	2.3	2.1
PCBM	--	4.3	--	1.8

A maximum value for the field effect mobility of holes in pristine MDMO-PPV, as reported from TFT investigation is around 10^{-5} cm^{2}V^{-1}s^{-1} [21].

PCBM is highly soluble in organic solvents, for example up to 40 mg/mL in chlorobenzene and 10 mg/mL in toluene. PCBM is an excellent electron acceptor, able to accept up to 6 electrons per molecule. The electron mobility in PCBM was investigated recently by employing several techniques and shows values in the range of 10^{-1} - 10^{-3} cm^{2}V^{-1}s^{-1}. However some reports even indicate ambipolar transport in PCBM [8, 22-26].

Figure 2.3 [20], shows the absorption coefficient of a PCBM film. According to the symmetry of the energy levels, the HOMO-LUMO transition is optically forbidden leading to a weak absorption at wavelengths above 500 nm.

2.1.2. Polymeric gate dielectrics

The most critical dielectric parameters that judge the usefulness of gate insulators are the following:

- the maximum possible electric displacement D_{max} the gate insulator can sustain [27]:

$$D_{max} = \varepsilon_0 \varepsilon E_B ,$$ (10)

where ε is the dielectric constant and E_B is the dielectric breakdown field;

- the capacitance per unit area C_i [27]:

$$C_i = \varepsilon_0 \left(\varepsilon \middle/ d_i \right) ,$$ (11)

where d_i is the insulator thickness. The magnitude of the capacitance per unit area is governed not only by the value of the dielectric function ε but also by the thickness, d_i, down to which pinhole-free films can be obtained, and thus also reflects the quality of the deposition technique in addition to intrinsic material properties [28].

Solution-processable materials are very attractive for applications in electronics, because films with sufficiently good insulator characteristics can often be formed by spin coating, casting or printing under ambient conditions at room temperature.

In this thesis, poly-vinyl-alcohol (PVA) and divinyltetramethyldisiloxane-bis(benzocyclobutene) (BCB) were chosen as gate-insulators, either because of the high dielectric constant (PVA) or because of the excellent dielectric properties (BCB). Both PVA and BCB deliver highly transparent films, an important prerequisite for optical active devices.

2.1.2.1. Poly-vinyl-alcohol (PVA)

Poly-vinyl-alcohol (PVA) is a water-soluble synthetic polymer with the chemical structure as shown in Fig. 2.4.

$$\underset{\displaystyle \text{PVA}}{\left[\begin{array}{c} \text{OH} \\ \ast \!-\!\!\!\!\diagdown\!\!\!\!\diagup\!\!\!\!\diagdown\!\!\!-\! \ast \end{array}\right]_n}$$

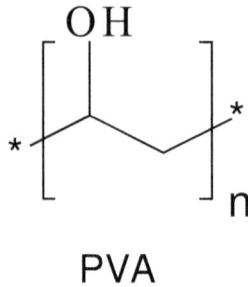

Fig. 2.4 Chemical structure of poly-vinyl-alcohol (PVA).

PVA has excellent film forming, emulsifying and adhesive properties. PVA is also resistant to oil, grease and to organic solvents. It is widely used in textile warp sizing, as an adhesive, as a paper sizing agent, as ceramic binder, and also in cosmetics as an emulsion stabilizer, in civil engineering and in the electronic industry.

PVA forms tough, clean, and transparent films. With a dielectric constant of 8 and charge electret properties (*electret*: material that has a permanent electric charge polarization), PVA could be a promising dielectric for applications in organic electronics [29, 30].

2.1.2.2. Divinyltetramethyldisiloxane-bis(benzocyclobutene) (BCB)

BCB belongs to a family of thermosetting polymer materials based on benzocyclobutene chemistry, developed by The Dow Chemical Company, which features simple processing, superior dielectric properties, excellent gap-fill properties and polarisation, low moisture absorption, and rapid cureing without the emission of reaction by-products [31]. The chemical structure of BCB is shown in Fig. 2.5.

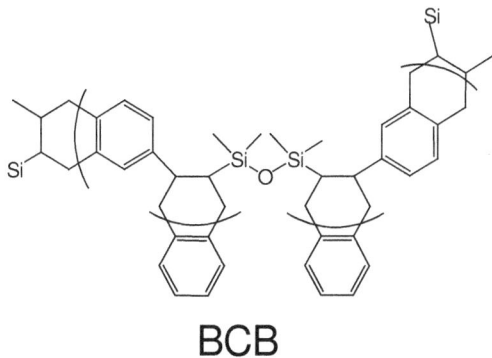

BCB

Fig. 2.5 Chemical structure of divinyltetramethyldisiloxane-bis(benzocyclobutene) (BCB).

The BCB monomer cures (crosslinks) thermally activated and forms a polymer network. A nearly complete cure can be obtained in a matter of minutes at 300° C. One hour is required at the usual cure temperature of 250° C [31].

The dielectric constant of BCB has negligible dispersion over a wide temperature and frequency range with a value close to 2.65 [32-34].

2.2. Device fabrication and characterisation

2.2.1 Device fabrication

Photoresponsive Organic Field Effect Transistors (photOFETs) are fabricated in top contact device geometry, as shown in Fig. 2.6. As substrates, glass sheets of 1.5 x 1.5 cm^2 covered with indium tin oxide ITO, from Merck KG Darmstadt, are used with an ITO thickness of ~ 120 nm and a sheet resistance < 15 Ωcm^{-2}.

The ITO is patterned by etching with an acid mixture of HCL_{konz}: HNO_{3konz}: H_2O (4.6:0.4:5) for ~ 15 min. The part of the substrate which forms the gate contact is coated with a commercial varnish to protect the active ITO layer against the etching acid. The varnish is removed after etching with acetone in an ultrasonic bath. The substrate is then cleaned in an ultrasonic bath with acetone and finally in iso-propanol.

The gate dielectrics are coated under ambient conditions on top of the cleaned and patterned ITO substrates by spin coating.

PVA with an average molecular weight of 127,000 (Sigma-Aldrich Mowiol®40-88) was used as received. PVA films were cast from a 10 % wt. aqueous solution by spin coating at

1500 rpm, yielding films with a thickness around 2 – 4 μm. The films were dried over night in an Argon atmosphere at 60°C. From the measured thickness of the dielectric layer, d and the dielectric constant $\varepsilon_{PVA} = 8$, a capacitance of $C_{PVA} = 1.8$ nF/cm^2 is estimated for the PVA based devices.

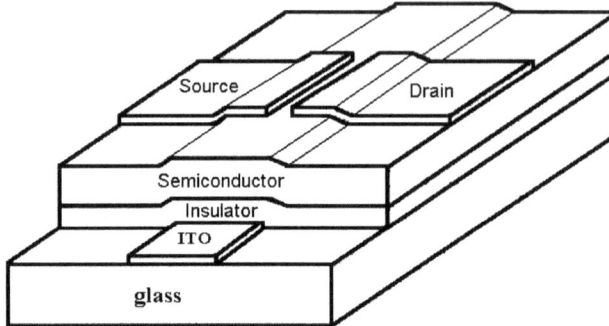

Fig. 2.6 Device structure of the top contact photOFET.

Divinyltetramethyldisiloxane-bis(benzocyclobutene) (BCB) (used as received from Dow Chemicals) was spin coated at 1500 rpm resulting in approximately 2-μm thick pinhole free films. The cross linking of BCB was carried out by curing at 250°C for 30 minute in a flowing Ar atmosphere inside a vacuum oven. BCB forms an inert dielectric layer with excellent mechanical properties and chemical stability, however with a rather low dielectric constant $\varepsilon_{BCB} = 2.6$.

As active material, a blend of MDMO-PPV (poly[2-methoxy-5-(3,7-dimethyloctyloxy)]-1,4-phenylenevinylene) (used as received from Covion) and (PCBM) methanofullerene [6,6]-phenyl C$_{61}$-butyric acid methyl ester (Solenne BV) (1 : 4 wt./wt. ratio) was spin coated inside a glove box under argon atmosphere at 1500 rpm for 40 s, followed by 2000 rpm for 30 s, from a 0.5 % chlorobenzene solution (1 % = 10 mg/ml), yielding films with a thickness around 170 nm.

As top source and drain contacts, LiF/Al (0.6/60 nm) were used, prepared by evaporation through a shadow mask under a vacuum of 5 x 10^{-6} mbar. Tungsten boats are used as deposition source. A quartz balance, Intellemetrics IC 600, monitors the deposition rate of the materials. The channel length, L, of the all devices was chosen to be 35 μm and the channel width, W, was 2 mm.

Non-volatile Organic Field-Effect Transistor Memory Elements based on a Polymeric Gate Electret was fabricated as follows: The device fabrication starts with the etching of the indium tin oxide (ITO) on the glass substrate. After patterning the ITO and cleaning in an ultrasonic bath, polyvinyl alcohol (PVA) was spin cast as soluble electret. PVA with a molecular weight of 100,000 was used as received from Fluka Chemicals. The PVA from Fluka was dissolved in distilled water and filtered using 0.2 μm filters, lyophilized and re-dissolved again in distilled water. With a 10 % wt. ratio of a highly viscous PVA solution a film thickness of 0.6 to 1 μm is achieved by spin coating at 1500 rpm. A methanofullerene [6,6]-phenyl C_{61}-butyric acid methyl ester (PCBM) active layer of 150 nm was spin coated on top of the PVA film from a chlorobenzene solution (3 wt %) in an argon atmosphere inside a glove box. The top source and drain electrodes, Cr (20 nm) were evaporated under vacuum (3 x 10^{-6} mbar) through a shadow mask. The schematic of the staggered mode non-volatile memory OFET is shown in Fig. 6.1 (Chapter 6).

Metal-Insulator-Metal (MIM) and *Metal-Insulator-Semiconductor* (MIS) devices are fabricated with the same procedures and with the same materials as the photOFETs. The devices are fabricated in a sandwich structure, as shown in Fig. 2.7. As gate-insulator, PVA (Sigma-Aldrich Mowiol®40-88) or BCB (Dow Chemicals) are used.

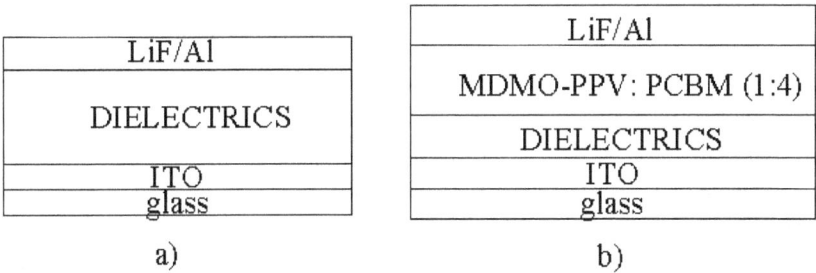

LiF/Al
DIELECTRICS
ITO
glass

a)

LiF/Al
MDMO-PPV: PCBM (1:4)
DIELECTRICS
ITO
glass

b)

Fig. 2.7 Schematic of a) Metal-Insulator-Metal (MIM) and Metal-Insulator-Semiconductor (MIS) devices.

2.2.2 Device characterisation

The electrical characterization was carried out under an inert argon environment inside a glove box system (MB 200 from Mbraun) or under vacuum. For the transistor characterisation, Keithley 2400 and Keithley 236 sourcemeter were used. OFET measurements have been performed with a scan rate of 2 Vs^{-1}.

The operation mode of the *OFET* is determined by the gate voltage, which can yield an accumulation layer of charges in the region of the conductive channel adjacent to the PVA/MDMO-PPV: PCBM (1:4) or BCB/MDMO-PPV: PCBM (1:4) interface, respectively.

For n-channel (or p-channel) operation mode, a positive (or negative) drain voltage is applied to induce an accumulation layer of electrons (or holes), allowing the measurement of the electron (or hole) mobility. The source contact was always connected to ground (see Fig.1.2).

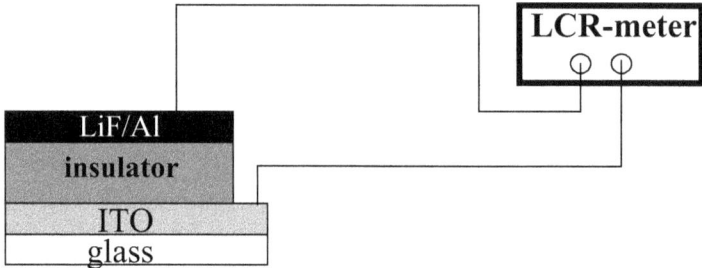

Fig. 2.8 *C-V* measurement set-up.

The capacity-voltage (C-V) characteristic is obtained with an impedance analyser by superimposing a small sinusoidal ac voltage signal to a DC voltage. By measuring the small-signal current, the phase and the modulus of the complex impedance can be extracted, and the capacity can be calculated. For the capacity-voltage *(C-V)* measurements we used a HP 4248A precision LCR meter with a typical scan rate of 0.2 Vs^{-1} in a frequency range from 1 Hz – 1 MHz (see Fig. 2.8). Devices were connected on the way that the high potential input (+) was applied to ITO side and the low potential input (-) to LiF/Al.

For the characterisation of the devices under light, a solar simulator (K.H. Steuernagel Lichttechnik GmbH) was used with light intensities ranging from 0.1 to 100 mW/cm^2 tuned by using neutral density filters. We have used this AM1.5 white light for two main reasons: *i)* it is a well defined and widely used standard, and *ii*) MDMO-PPV: PCBM blend based solar cells have been intensively investigated under this very light, in our group [1, 3, 19].

As a monochromatic light source, an Ar^+ laser, INNOVA 400, with a wavelength of 514 nm is used, typically with an optical power of 120 mW. The illumination intensity was varied by using neutral density filters. The illuminated spot on the sample has an area around 4 mm^2.

Devices were illuminated through the ITO coated glass and through the transparent dielectric layer.

The surface morphology and the thickness of the dielectrics and of the active layers were measured under ambient conditions with a Digital Instruments Dimension 3100 atomic force microscope in the tapping mode.

For the *spectral characterisation* of the BCB-photOFETs, the following experiment was employed; The BCB-photOFET was prepared as described above (see § 2.2.1). The sample was loaded in a chamber purged a constant nitrogen flow. A glass window in the front of the chamber allows illumination of the device, as shown in Fig. 2.9. The device was illuminated through the transparent ITO gate and dielectric. As light source, a 900W Xenon lamp was used. The output radiation of the lamp was passed through a monochromator with a spectral resolution of $d\lambda < 2$ nm and focused on the sample, so that the device channel was fully illuminated. The monochromator slits were kept at 200 µm. The wavelengths were changed in steps of 10 nm, from 800 to 350 nm. During the measurement, the incident light power was kept constant at 0.375 mW by regulating the lamp current. A calibrated silicon photodiode was used for monitoring the incident light power. Constant illumination intensity was required because the photogenerated current does not increase linearly with the illumination intensity at high gate voltages and therefore cannot be normalized. For the transistor characterisation, Keithley 2400 and Keithley 236 sourcemeters were used. OFET measurements were performed with a scan rate of 2 Vs^{-1}. By recording the transistor transfer characteristics at selected wavelengths and by making measurements at certain V_{gs} values (constant $V_{ds} = 80$V), the photocurrent spectrum is obtained (see Fig. 5.22, Chapter 5).

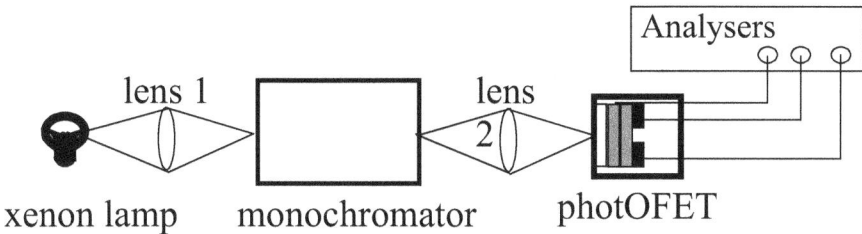

Fig. 2.9 Optoelectronical set-up.

2.3 References

[1] S.E. Shaheen, C.J. Brabec, N.S. Sariciftci, F. Padinger, T. Fromherz, and J.C. Hummelen, *Appl. Phys. Lett.* 78, 841, 2001.

[2] V. Dyakonov, *Physica E*, 14, 53, 2002.

[3] C.J. Brabec, N.S. Sariciftci, and J.C. Hummelen, *Adv. Func. Mater.* 11, 15, 2001.

[4] C.J. Brabec, A. Cravino, D. Meissner, N.S. Sariciftci, T. Fromherz, T. Rispens, L. Sanchez, and J.C. Hummelen, *Adv. Func. Mater.* 11, 374, 2001.

[5] J.K.van Duren, J. Loos, F. Morissey, C.M. Leewis, K.P.K. Kivits, L.J. van Ijzendoorn, M.T. Rispens, J.C. Hummelen, and R.A.J. Janssen, *Adv. Func. Mater.* 12, 665, 2002.

[6] J.K. Kroon, M.M. Wienk, W.J.H. Verhees, and J.C. Hummelen, *Thin Solid Films*, 403-404, 223, 2002.

[7] M.Al-Ibrahim, H.K. Roth, and S. Sensfuss, *Appl. Phys. Lett.* 85, 1481, 2004.

[8] V.D. Mihaliletchi, J.K.J. van Duren, P.W.M. Blom, J.C. Hummelen R.A.J. Janssen, J.M. Kroon, M.T. Rispens, W.J.H. Verhess, and M.M. Wienk, *Adv. Func. Mater.* 13, 43, 2003.

[9] J.K.J. van Duren, V.D. Mihailetchi, P.W.M. Blom, T. van Woudenbergh, J.C. Hummelen, M.T. Rispen, R.A.J. Janssen, and M.M. Wienk, *Appl. Phys. Lett.* 94, 4477, 2003.

[10] E. J. Meier, D. M. de Leeuw, S. Setayesh, E. van Veenendaal, B. -H. Huisman, P. W. M. Blom, J. C. Hummelen, U. Scherf, and T. M. Klapwijk, *Nature Mater.* 2, 678, 2003.

[11] S.A. Choulis, J. Nelson, Y. Kim, D. Poplavskyy, J.R. Durrant, and D.D.C. Bradley, *Appl. Phys. Lett.* 83, 2003

[12] C. Maler, E.J. Koop, D. Mihailetchi, and P.W.M. Blom, *Adv. Func. Mater.* 14, 865, 2004.

[13] T.J. Savenije, J.E. Ktoeze, M.M. Wienk, J.M. Kroon, and J.M. Warman, *Phys. Rev. B*, 69, 155205, 2004.

[14] A.J. Mozer, G. Dennler, N.S. Sariciftci, M. Westerling, A. Pivrikas, R. Österbacka, and G. Juška, *Phys. Rev. B*, 72, 035217, 2005.

[15] W. Geens, T. Martens, J. Poortmans, T. Aernouts, J. Manca, L. Lutsen, P. Heremans, S. Borghs, R. Mertens, and D. Vanderzande, *Thin Solid Films*, 451-452, 498-502, 2004.

[16] Th. B. Singh, S. Günes, N. Marjanović, N. S. Sariciftci, and R. Menon, *J. Appl. Phys.* 97 114508, 2005.

[17] H. Spreitzer, H. Becker, E. Kluge, W. Kreuder, H. Schenk, R. Demandt, and H. Schoo, *Adv. Mater.* 10, 1340, 1998.

[18] J.C. Hummelen, B.W. Knight, F. LePeq, F. Wudl, J. Yao, and C.L. Wilkins, *Journ. Org. Chem.* 60, 532, 1995.

[19] C. Winder, *PhD Thesis*, JKU Linz, 2004.

[20] H. Hoppe, N. Arnold, N.S. Sariciftci, and D. Meissner, *Solar Energy Materials & Solar Cells*, 80, 105 – 113, 2003.

[21] W. Geens, S.E. Shaheen, B. Wessling, C.J. Brabec, J. Poortmans, and N.S. Sariciftci, *Org. Electr.* 3, 105-110, 2002.

[22] Th. B. Singh, N. Marjanović, P. Stadler, M. Auinger, G. J. Matt, S. Günes, N. S. Sariciftci, R. Schwödiauer, and S. Bauer, *J. Appl. Phys.* 97, 083714, 2005.

[23] C. Waldauf, P.Schilinsky, M. Perisutti, J. Hauch, and C.J. Brabec, *Adv. Mater.* 15, 2081, 2003.

[24] R. Pacios, J. Nelson, D.D.C. Bradley, and C.J. Brabec, *Appl. Phys.Lett.* 83, 4764, 2003.

[25] T.D. Anhtopoulos, D.M. de Leeuw, E. Cantatore, S. Setayesh, E.J. Meijer, C. Tanase, J.C. Hummelen, and P.W.M. Blom, *Appl. Phys. Lett.* 85, 4205, 2004.

[26] T.D. Anhtopoulos, C. Tanase, S. Setayesh, E.J. Meijer, J.C. Hummelen, P.W.M. Blom, and D.M. de Leeuw, *Adv. Mater.* 16, 2174, 2004.

[27] A. Facchetti, M.-H. Yoon, and T. Marks, *Adv. Mater.* 17, 1705, 2005.

[28] J. Vares, S.D. Ogier, S.W. Leeming, D.C. Cupertino, and S.M. Khaffaf, *Adv. Func. Mater.* 13, 199, 2003.

[29] Th. B. Singh, N. Marjanović, G. J. Matt, N. S. Sariciftci, R. Schwödiauer, and S. Bauer, *Appl. Phys. Lett,* 85, 5409, 2004.

[30] Th. B. Singh, F. Meghdadi, S. Günes, N. Marjanović, G. Horowitz, P. Lang, S. Bauer, and N.S. Sariciftci, *Adv. Mater.* 17, 2315, 2005.

[31] M.E. Mills, P. Townsend, D. Castillo, S. Martin, and A. Achen, *Microelectronic Engineering,* 33, 327-334, 1997.

[32] L. -L. Chua, P. K. H. Ho, H. Sirringhaus, and R. H. Friend, *Appl. Phys. Lett.* 84, 3400, 2004.

[33] R. Schwödiauer, G. S. Neugschwandtner, S. Bauer-Gogonea, S. Bauer, and W. Wirges, *Appl. Phys. Lett.* 75, 3998, 1999.

[34] Th. B. Singh, N. Marjanović, G. J. Matt, S. Günes, N. S. Sariciftci, A. M. Ramil, A. Andreev, H. Sitter, R. Schwödiauer, and S. Bauer, *Org. Electr.* 6, 105, 2005.

Chapter 3

3. *I-V* and *C-V* characterisation of MDMO-PPV: PCBM (1:4) blend based diodes

3.1. Conjugated polymer/fullerene blend based diodes

Among the methods employed for the electrical characterisation of semiconductors, the steady state current-voltage (*I-V*) and the dynamic capacitance-voltage (*C-V*) techniques are the most widely used. Carrier mobility, carrier traps, contact barrier height, surface states, are among the electrical parameters that can be determined by these methods.

3.1.1 *I-V* characteristics of the MDMO-PPV: PCBM (1:4) blend based diodes

The glass/ITO/MDMO-PPV: PCBM (1:4)/LiF-Al diode is characterised under dark and under illumination, as shown in Fig. 3.1.

The *I-V* curve in the dark shows a typical diode characteristic with a good rectification ratio of about 10^6 at +/- 5V. The device was illuminated with selected monochromatic light, with a wavelength of $\lambda = 514$ nm, which matches with the maximum of the absorption peak of the MDMO-PPV. The illumination intensities chosen are 6 mWcm^{-2} and 30 mWcm^{-2}. Under the illumination with intensity of 6 mW.cm^{-2} a photovoltaic effect is seen. A significant increase of the photocurrent in reverse bias is observed due to the illumination of the device. The increase in the photocurrent can be attributed to the photogeneration of free charge carriers in the highly photoactive blend upon illumination. As expected, an increase of the illumination intensity to 30 mW.cm^{-2} induces a further increase of the photocurrent especially under reverse bias.

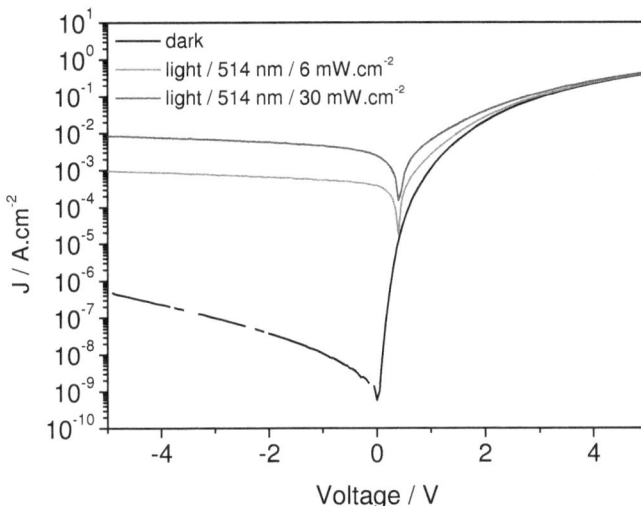

Fig. 3.1 *I-V* characteristics of the MDMO-PPV: PCBM (1:4) blend based diodes in the dark and under monochromatic illumination.

3.1.2 C-*V* characteristics of MDMO-PPV: PCBM (1:4) blend based diodes

In addition to the *I-V* curves also capacitance-voltage measurements were performed. The scheme of the measurement set-up has already been discussed in Fig. 2.8 of Chapter 2.

Capacitance-voltage measurement were done in the dark and under selected illumination conditions (λ = 514 nm, 6 mW.cm^{-2} and 30 mWcm^{-2}) as in the case of the *I-V* measurements. The frequency range investigated covers almost 5 decades (40 Hz to 1 MHz). From the *C-V* measurements a blend thickness of ~ 200 nm was calculated by using equation 11 (Chapter 2). The measurement results are shown in Figs. 3.2 – 3.6.

Fig. 3.2 shows the *C-V* characteristics of the MDMO-PPV: PCBM (1:4) blend in the dark and under illumination. For a measurement frequency of 1 kHz in the dark, the capacitance is constant when the diode is biased in the reverse direction. However, when the charge injection starts under forward bias, the capacitance strongly increases. The capacitance saturates at +2.7 V, before it decreases even to negative values at high forward bias voltages. Similar variations of the capacitance in organic semiconductor diodes were reported in the literature. Explanations for the experimentally obtained results are based on different mechanisms [1-6]. The constant capacitance under reverse bias clearly indicates that the semiconductor is entirely depleted: The presence of a Schottky type contact can be ruled out,

27

and the capacitance measured equals the geometrical capacitance. Under forward bias, charge injection induces a drastic increase of the capacitance. It has been proposed that saturation occurs when double injection starts, and that the following decrease takes its origin in the recombination of charges [1].

Under illumination (514 nm, 6 mW.cm^{-2}), the C-V characteristic depicted in Fig. 3.2 was observed (red curve). However, despite the similar values observed under reverse bias, the value of the capacitance is considerably enhanced under forward bias. This result sounds counterintuitive at first sight: Photogenerated charge carriers are expected to be detected in the reverse direction, while photoconductivity, almost absent in the graph displayed in Fig 3.1, is not expected to modify the value of the capacitance. Further investigations under reverse bias, as detailed below have been undertaken to clarify the experimentally obtained results. It seems justified to add a comment here concerning the negative values of the capacitance observed at large forward bias. The "negative capacitance" in organic electronic devices is attracting considerable attention [7-12]. The interest arises from the fact that a negative capacitance might open new application routes, like compensation circuits. However, one has to be very careful: The negative values are calculated by the set-up using equivalent circuits from the measured modulus and phase of the complex impedance. Therefore it seems more appropriate to display the complex impedance of the device (which is not dependent on equivalent circuits) in the representation of experimental results. The negative capacitance under large forward bias voltage is interpreted in terms of space charge limitations of the generated charge carriers, of recombination kinetics, etc [11, 12]. No clear picture is presently available which can be used to model measured results in a convincing way.

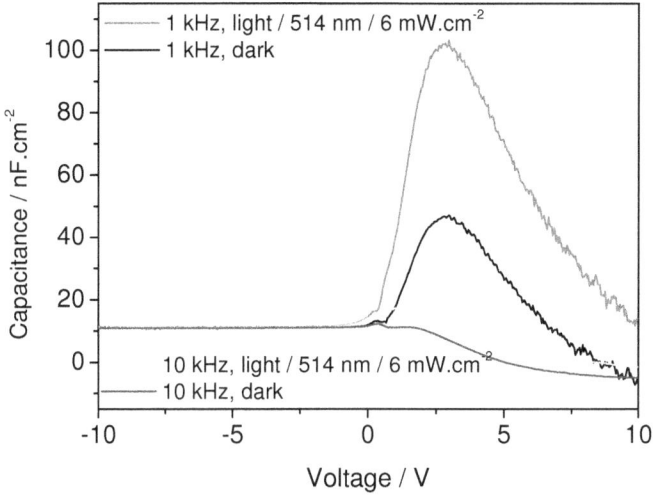

Fig. 3.2 *C-V* characteristics of MDMO-PPV: PCBM (1:4) blend based diodes in the dark and under monochromatic illumination at frequencies of 1 and 10 kHz.

The capacitance vs. frequency of the MDMO-PPV: PCBM (1:4) blend based diode at given bias voltages of 0 V and –5 V in the dark and under illumination is plotted in Fig. 3.3. One should note that the decrease in the capacitance values at frequencies above 600 kHz might be caused by the connection of the sample to the impedance analyser, which becomes always important at high frequencies. Therefore, only results for frequencies below 500 kHz are discussed here. In the dark at –5 V, the capacitance appears to be frequency independent. This is consistent with the statement made above: The capacitance measured under reverse bias is the geometrical capacitance of a sample consisting of two electrodes separated by a totally depleted organic semiconductor. On the other hand, the frequency dependence of the capacitance under forward bias indicates that charge carriers are present in the device, and that the mobility of these carriers is limited: Their contribution to the capacitance fades off above 10 kHz, showing that the carriers cannot follow such fast oscillating ac-fields.

In order to investigate more precisely the reverse bias regime, more measurements were performed, as shown in Figure 3.4.

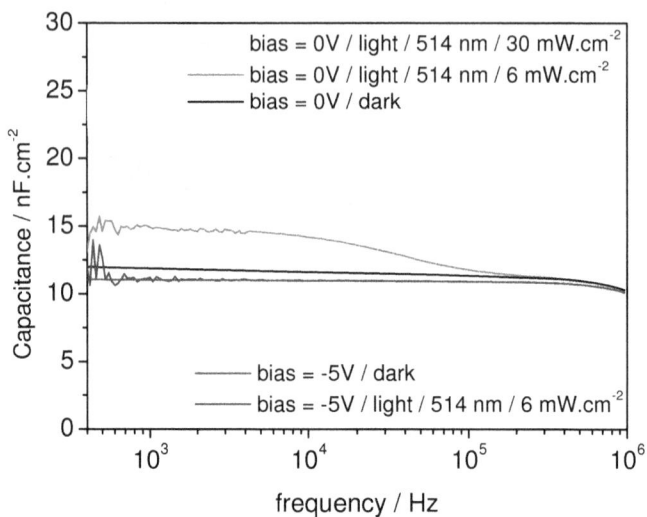

Fig. 3.3 Capacitance vs. frequency of the MDMO-PPV: PCBM (1:4) diodes in the dark and under illumination at bias voltages of 0 V and –5 V bias.

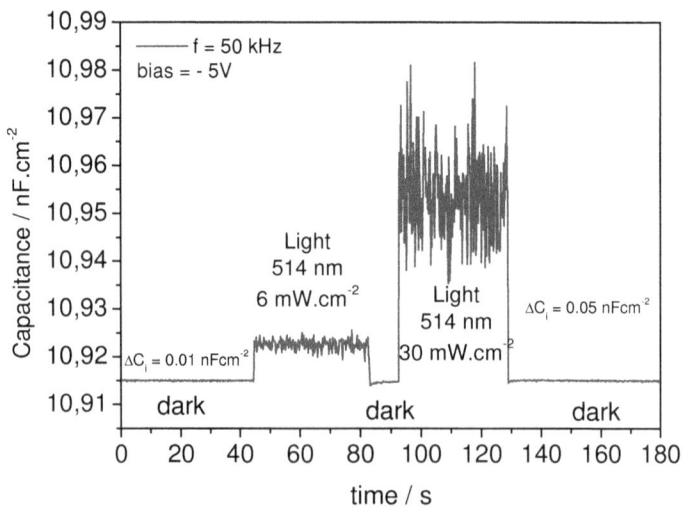

Fig. 3.4 Transient capacitance under reverse bias (–5 V) in the dark and under illumination.

The device was biased at – 5 V. A frequency of 50 kHz was selected in order to avoid the noise visible at low frequencies in Fig 3.4. The device was successively illuminated with different illumination intensities. A relatively small increase in the capacitance upon illumination with respect to the dark value is observed. The increase in the capacitance can be directly correlated to the presence of photoinduced charge carriers since the capacitance increase upon illumination ΔC_i, is directly proportional to the ratio of the illumination intensities.

Transient capacitance measurements at zero bias are shown in Fig. 3.5. Here, the device was illuminated with one wavelength and one illumination intensity, while the frequency was varied from 1 kHz to 100 kHz. The capacitance increase upon illumination ΔC_i, is higher than in the previous case when the device is biased at –5 V. In addition the capacitance increase is also frequency dependent. The results obtained here are consistent with the measurements shown in Fig. 3.2: The light induced modification of the capacitance is much larger under forward bias as compared with the reverse bias operation.

In order to verify the existence of potentially long living trapped charges, we investigated the capacitance decay when the light is switched off. Typical results are plotted in Figure 3.6: The capacitance evolves exponentially, needing more than 300 seconds recovering the steady state value. The experiment points out that long-living trapped charges exist in the blend, though their overall number might be quite small.

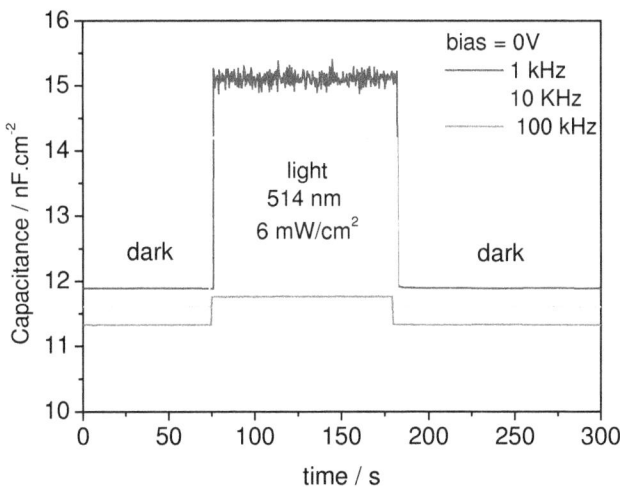

Fig. 3.5 Transient capacitance at 0 V bias, under different illumination conditions.

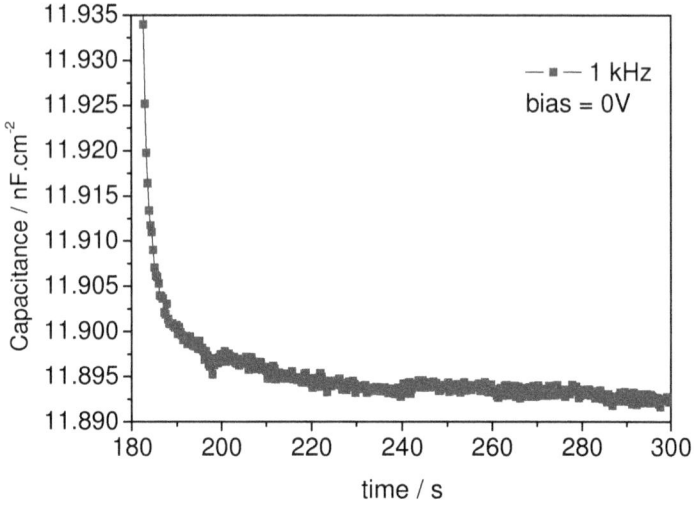

Fig. 3.6 Transient capacitance of an MDMO-PPV: PCBM (1:4) diode after switching off the light illumination.

3.2 Summary

In summary, we investigated MDMO-PPV: PCBM (1:4) blend based diodes with I-V and C-V characterisations. The I-V characteristics show typical diode behaviour in the dark whereas upon illumination an increase in the photocurrent was observed especially under reverse bias. This increase in the current can be attributed to the photogeneration of charge carriers in the highly photoactive blend upon illumination.

Capacitance-voltage measurements were performed in the dark and under selected illumination conditions, similar to those used for the I-V measurements. The existence of a "negative capacitance" at high injecting bias voltage was observed. Explanation for this phenomenon are still not quite clear yet, although several possible scenarios were proposed in the literature (*i.e.* intrinsic property of disorder materials [7]; phase shift of the impedance due to space charge limited current [10, 11], etc).

In reverse bias only a small change in the capacitance was observed under illumination. Nevertheless, the small increase in the capacitance can be directly correlated with the presence of photoinduced charge carriers since the capacitance increase under illumination, ΔC_i, is directly proportional to the illumination intensities. Finally, a presence of a small number of long-living trapped charge carriers was observed.

3.3 References

[1] V. Shrotriya and Y. Yang, *J. Appl. Phys*. 97, 054504, 2005.

[2] I.H. Campbell, D.L. Smith, C.J. Neef, and J.P. Ferraris, *Appl. Phys. Lett*. 72, 2565, 1998.

[3] V. Dyakonov, D. Godovsky, J. Mayer, J. Parisi, C.J. Brabec, N.S. Sariciftci, and J.C. Hummelen, *Synth. Met*. 124, 103-105, 2001.

[4] I.H. Campbell, D.L. Smith, and J.P. Ferraris, *Appl. Phys. Lett*. 66, 3030, 1995.

[5] W. Brütting, H. Riel, T. Beierlein, and W. Riess, *J. Appl. Phys*. 89, 1704, 2001.

[6] W. Riess, H. Riel, T. Beierlein, W. Brütting, P. Müller, and P.F. Seidler, *IBM J. RES. & DEV*. 45, 77, 2001.

[7] H.L. Kwok, *Solid-State Electronics*, 47, 1089, 2003.

[8] F. Lemmi and N.M. Johnson, *Appl. Phys. Lett*. 74, 251, 1999.

[9] A.G.U. Perera, W.Z. Shen, M. Ershov, H.C. Liu, M. Buchanan, and W.J. Schaff, *Appl. Phys. Lett*. 74, 3167, 1999.

[10] H.C.F. Martens, W.F. Pasveer, H.B. Brom, J.N. Huiberts, and P.W.M. Blom, *Phys. Rev. B*, 63, 125328, 2001.

[11] H.C.F. Martens, H.B. Brom, and P.W.M. Blom, *Phys. Rev. B*, 60, R8489, 1999.

[12] L.S.C. Pingree, B.J. Scott, M.T. Russell, T.J. Marks, and M.C. Hersam, *Appl. Phys. Lett*. 86, 073509, 2005.

Chapter 4

4. photOFETs based on MDMO-PPV: PCBM (1:4) blends on top of a PVA gate-insulator

Capacitance-voltage (C-V) measurements can be used to determine charges in insulators like charges located in interface traps, bulk traps, as well as mobile ionic charges. Hence capacitance-voltage measurements are widely used in order to gain information on the electrical properties of insulating materials or on interface effects. Therefore, prior to the presentation of the transistor characteristics photOFETs based on PVA gate insulators and MDMO-PPV: PCBM (1:4) blends as the photoactive semiconductor layer, C-V characteristics of MIM or MIS structure are discussed.

4.1 PVA based MIM device

Metal-Insulator-Metal (MIM) devices based on pristine PVA dielectrics were prepared and characterised.

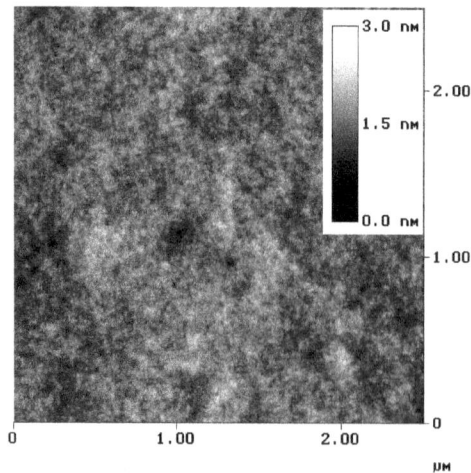

Fig. 4.1 AFM image of a PVA dielectric film.

Fig. 4.1 shows the height image of spin coated PVA dielectric films obtained by AFM measurements in the tapping mode. It can be concluded from the figure that the PVA dielectric provides a smooth surface with a roughness below 3 nm.

The capacitance vs. frequency of PVA based MIM devices in the dark are shown in Fig. 4.2. The capacitance strongly depends on the measurement frequency. The capacitance increase with decreasing frequency may be explained on the basis of charge carriers being blocked at the electrodes. However, an alternative explanation might be given in terms of molecular dipoles presented in PVA that cannot follow the applied electric field at large frequencies [1]. Without additional measurements, performed over a wide range of temperatures, these dielectric measurements cannot be unanimously analysed.

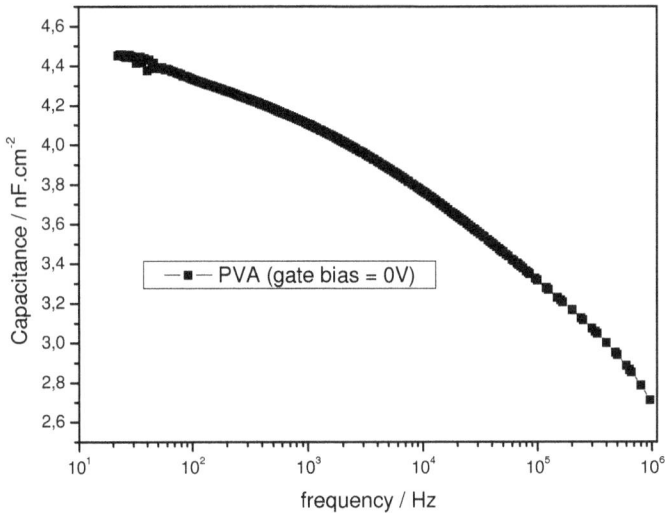

Fig. 4.2 Capacitance vs. frequency of PVA based MIM devices.

4.2. PVA/MDMO-PPV: PCBM (1:4) blend based MIS devices

4.2.1. Dark condition

Metal-Insulator-Semiconductor (MIS) devices based on PVA and on the MDMO-PPV: PCBM (1:4) blend as active layer were fabricated and characterised in the dark and under illumination (as described earlier, Chapter 2). As top source and drain contacts, LiF/Al was used.

The surface properties of the blend films on PVA were measured with an AFM (Fig. 4.3).

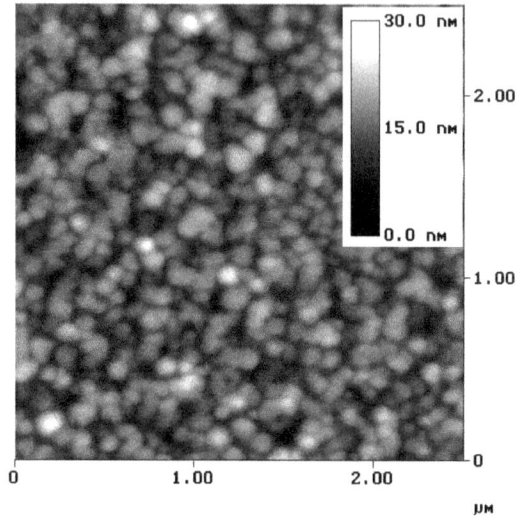

Fig. 4.3 AFM image of a PVA/MDMO-PPV: PCBM (1:4) blend film.

The blend films coated on top of the PVA dielectric show a roughness below 25 nm. A phase separation as already observed earlier is also noted [2, 3].

The capacitance vs. frequency of PVA/MDMO-PPV: PCBM (1:4) blend based MIS device in the dark is shown in Fig. 4.4. With a negative bias voltage, no significant change in the capacitance over a wide range of frequencies was observed. However, when the device is biased with a positive voltage, injected electrons, which are accumulated at the PVA/blend interface, induce an increase of the capacitance.

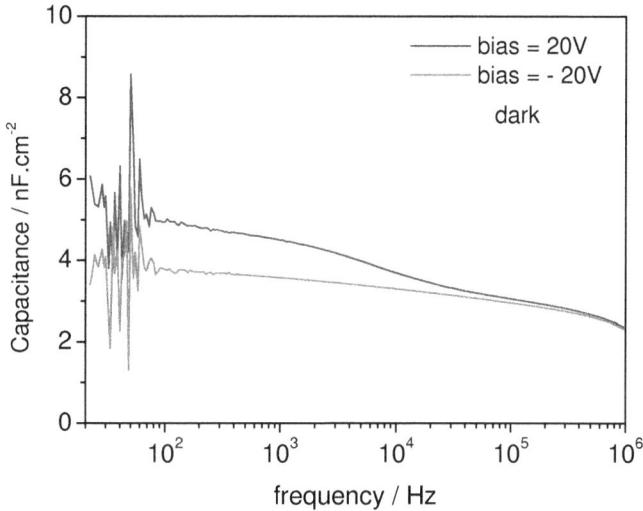

Fig. 4.4 Capacitance vs. frequency of PVA/MDMO-PPV: PCBM (1:4) blend based MIS devices.

Since the blend is composed of a mixture of p- and n-type semiconductors, one could expect injection/accumulation of both charge carriers in the blend. In the case of LiF/Al top contacts, only electron injection/accumulation was observed.

C-V characteristics of the same device in the dark are shown in Fig. 4.5. The voltage was scanned from –40 V up to 40 V in the forward and back direction at a given frequency. At the lowest frequency (333 Hz), starting from –40 V up to –7 V no significant injection of holes is observed (Fig. 4.5, upper curves). Massive electron injection started beyond –7 V, inducing an increase in the capacitance. Approximately between +20 V up to +40 V the capacitance saturated. In order to bring the capacitance back to the initial value, a large negative bias voltage must be applied. As a result, a significant hysteresis in the C-V curves occurs. Similar C-V characteristics are observed at high frequencies, yet the difference between the maximum and the minimum capacitance does decrease with increasing frequency.

The results suggest that charge carriers, which are trapped/detrapped at the dielectric/blend interface, are responsible for the occurrence of a hysteresis in the C-V characteristics.

In other systems, the occurrence of hysteresis loops has been attributed to charge trapping effects at the semiconductor/dielectric interface or in the bulk of the dielectric as well [4-6].

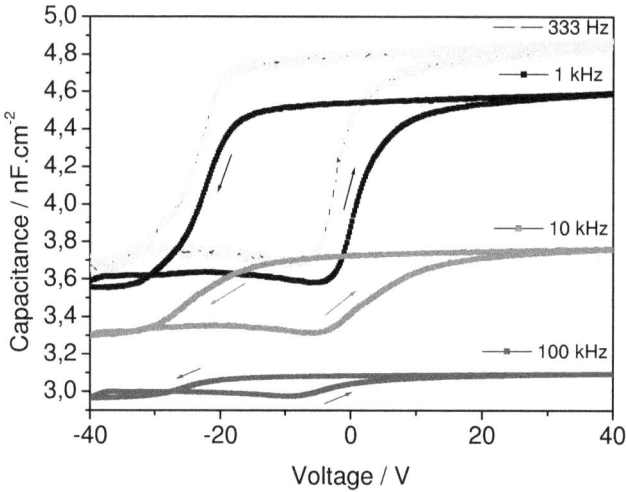

Fig. 4.5 Capacitance-Voltage characteristics of the PVA/MDMO-PPV: PCBM (1:4) blend based MIS devices in the dark. The arrows show the sweep directions, starting at – 40 V

Another interesting observation is that the total capacitance of the system is slightly higher than the total capacitance of the MIM device based on pristine PVA dielectrics. This effect is not yet fully understood.

In the classical picture the total capacitance of capacitors in series is given by:

$$\frac{1}{C_t} = \frac{1}{C_i} + \frac{1}{C_s},$$
(12)

where C_t is the total capacitance, C_i the insulator capacitance and C_s the semiconductor capacitance, respectively.

According to equation (12), the maximum value of C_t should be the insulator capacitance. The observed results indicate a negative capacitance of the semiconductor. However, other effects might also induce the unexpected higher value of C_t, like for example parasitic parallel capacitances due to the geometry of the device.

4.2.2 Under monochromatic illumination

The same device was characterised in the dark and under monochromatic conditions. As light source an Ar$^+$ laser with a wavelength of 514 nm and an intensity of 6 mW.cm^{-2} was used. The transient capacitance was recorded in the dark and then under monochromatic illumination and finally in the dark again after illumination at a given frequency and light intensity (Fig. 4.6). The capacitance shows a voltage dependence, with a larger light to dark ratio in the case of negative bias voltage. In this case, the device is depleted of charge carriers and the light effect is more pronounced. In the case of positive bias (accumulation mode), electrons are injected by the contact and the contribution of the light induced charges to the total capacitance is less visible.

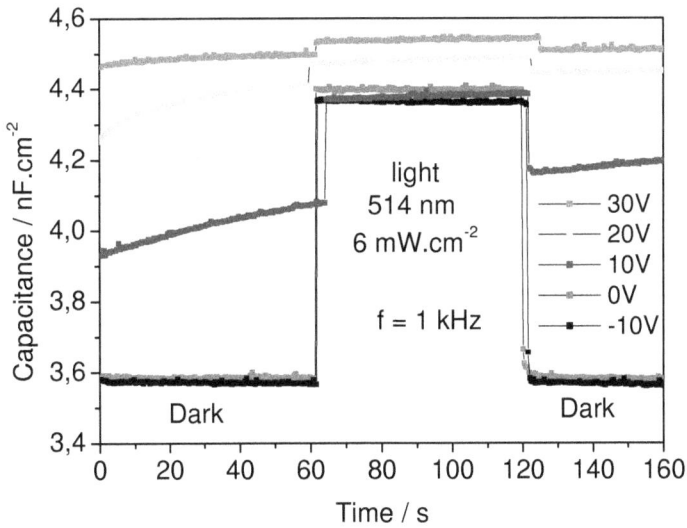

Fig. 4.6 Transient capacitance of the PVA/MDMO-PPV: PCBM (1:4) blend based MIS devices in the dark and under monochromatic illumination.

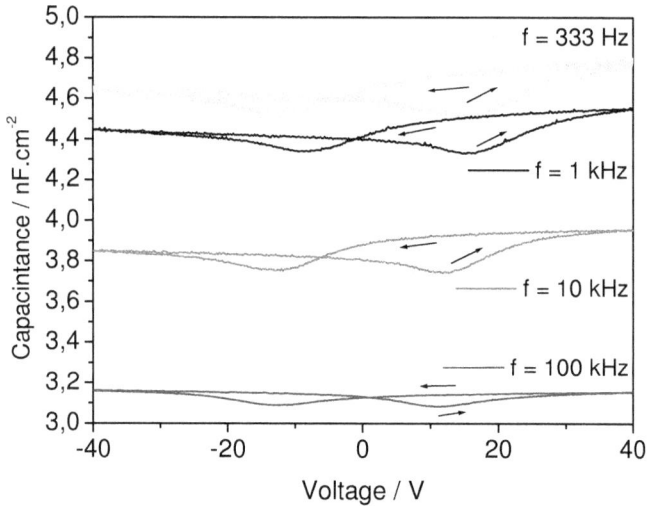

Fig. 4.7 *C-V* characteristics of the same device under monochromatic illumination. The arrows indicate the sweep directions, starting at – 40 V.

Fig. 4.7 shows the *C-V* characteristics under illumination for the following illumination conditions: λ = 514 nm, 6 mW.cm^{-2}. Arrows are used to indicate the sweep direction. A significant increase in the capacitance as compared to the values in the dark is observed under negative bias voltage (see Fig. 4.5). A general trend of the *C-V* characteristics is that the threshold voltage shifts to higher positive voltages in comparison to the dark, presumably due to complex interactions of different effects like: charge trapping at the PVA/blend interface or in the bulk of the dielectric or additional electric field induced charges.

C-V characteristics in the dark, under illumination and in the dark after illumination are depicted in Fig. 4.8. *C-V* characteristics taken in the dark after illumination had a similar initial hysteresis shape. The threshold voltage shift obtained under illumination can be recognised in the shape of the *C-V* characteristic taken in the dark after illumination. Again, a possible reason for this effect can be charge-trapping effects at the PVA/blend interface.

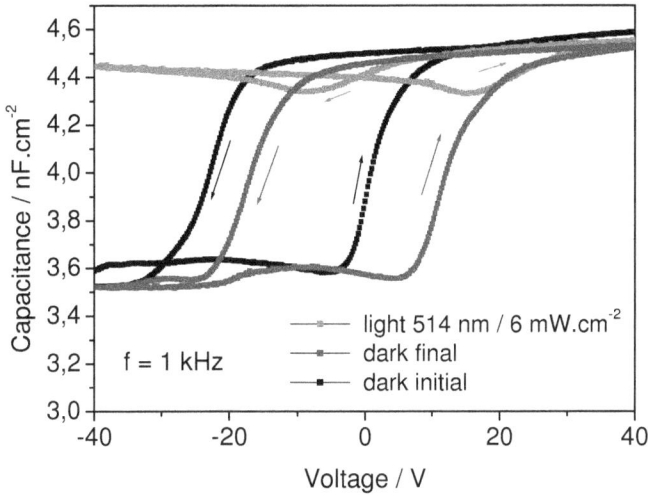

Fig. 4.8 Capacitance vs. voltage of the PVA/MDMO-PPV: PCBM (1:4) blend based MIS devices in the dark, under illumination and in the dark after illumination. The arrows indicate the sweep direction, starting at – 40 V.

4.2.3. Under AM1.5 illumination

C-V characteristics are taken under white light illumination conditions (AM1.5/100 mW.cm^{-2}), and shown in Fig. 4.9. As white light source solar simulator was used. Significant changes in the C-V characteristics occur upon illumination, as depicted in Fig. 4.9. At the beginning of the measurement, an enormous increase in the capacitance was observed, followed with by a strong decrease to a stable value, also in the return scan direction (from +40 V to –40 V).

Fig. 4.10 shows C-V characteristics of the same device in the dark after illumination. As depicted, no further voltage dependence was observed, as explained by the absence of an accumulation or depletion-operating mode.

The observed behaviour is explained as a severe photoinstability of the system and corresponding device degradation. It should also be noted that during illumination the device is subject to significant heating. The PVA itself is a transparent dielectric polymer which is not expected to yield a complete photodegradation within the short measurement times of the experiment. Furthermore, the MDMO-PPV/PCBM mixture used here is a well known solar cell material and has been shown to be reasonably stable in a glove box under a controlled

Argon atmosphere. Therefore, the degradation observed here (in Fig. 4.9) is probably due to the interface degradation between the PVA dielectric and the photoactive layer. This is also supported by the fact that devices based on the other dielectric used in this thesis (BCB) reveals a much more stable capacitance response upon illumination. Still, the origin of the processes and mechanisms, which caused the photoinstability and degradation, are not quite clear yet.

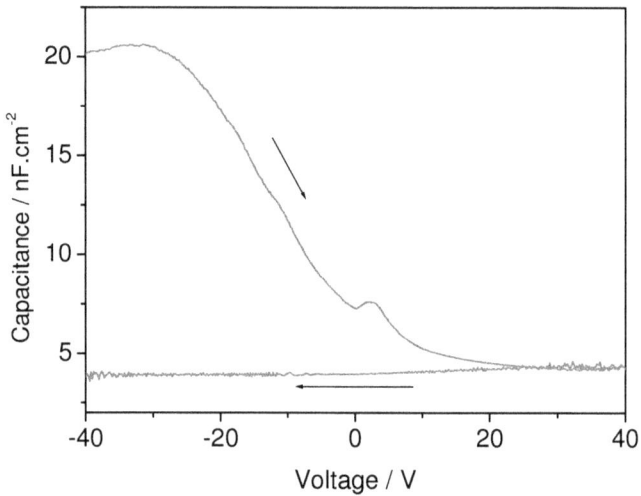

Fig. 4.9 Capacitance-Voltage characteristics of the PVA/MDMO-PPV: PCBM (1:4) blend based MSM device under white light illumination condition (AM1.5 /100 mW.cm^{-2}). The arrows indicate the sweep direction, starting at – 40 V.

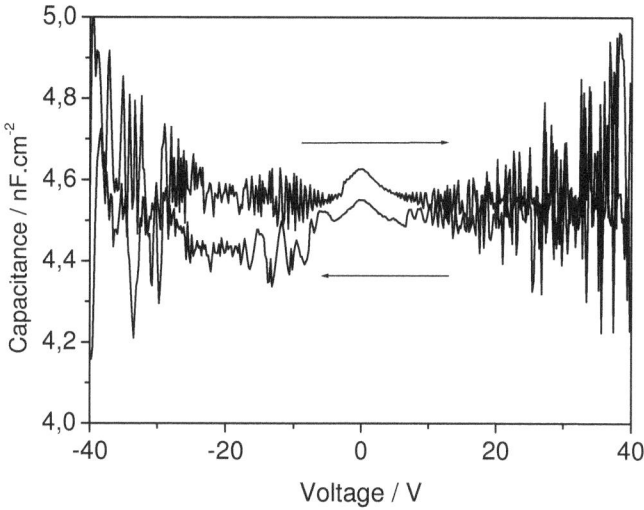

Fig. 4.10 Capacitance-Voltage characteristics of the PVA/MDMO-PPV: PCBM (1:4) blend based MSM device in dark post-white light illumination at 1 kHz. The arrows indicate the sweep direction, starting at – 40 V.

4.3. PVA/MDMO-PPV: PCBM (1:4) blend based photOFETs

Photoresponsive Organic Field-Effect Transistors (photOFETs) based on PVA and MDMO-PPV: PCBM (1:4) blends (PVA-photOFETs) are fabricated and characterised in the dark, under monochromatic illumination and under white light conditions (AM1.5). For the top source and drain contacts LiF/Al is used. The device fabrication and the characterisation techniques used are described in Chapter 2.

4.3.1. Dark condition

The output characteristics of MDMO-PPV: PCBM (1:4) - PVA photOFETs with LiF/Al top source and drain contacts in the dark are shown in Fig. 4.11. An electron accumulation mode is achieved with a positive bias gate voltage, V_{gs}, demonstrating n-type transistor behaviour, similar to the behaviour reported in pristine PCBM based devices [4]. It is assumed that LiF/Al forms ohmic contacts with the blend layer, especially with respect to charge injection and collection from the fullerene phase [7, 8].

Fig. 4.12(a) and 4.12(b) shows transfer characteristics and square root of the source-drain current at drain-source voltage V_{ds} = 80 V in the dark, respectively. An electron field-effect mobility, μ, of 10^{-2} cm^2/Vs was calculated from the saturation regime by using equation (2) (see Chapter 1). In pristine PCBM based OFETs, calculated electron mobilities as high as 0.2 cm^2/Vs have been observed [4], at least one order of magnitude larger than in the present blend devices. In both cases the device geometry, dielectric and metal contact are similar. Therefore, we presume that the polymer chains significantly disturb the inter-molecular hopping transport in the fullerenes phase of the blend, lowering its electron field-effect mobility. Again, by changing the sweep direction, a hysteresis was observed in the transfer curve, as in the case of the PVA/blend based MIS device.

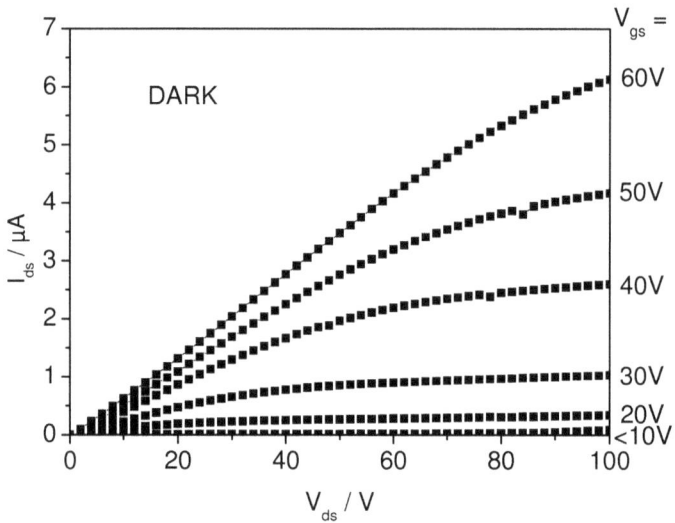

Fig. 4.11 Output characteristics of the PVA-photOFETs with LiF/Al source and drain contacts in the dark.

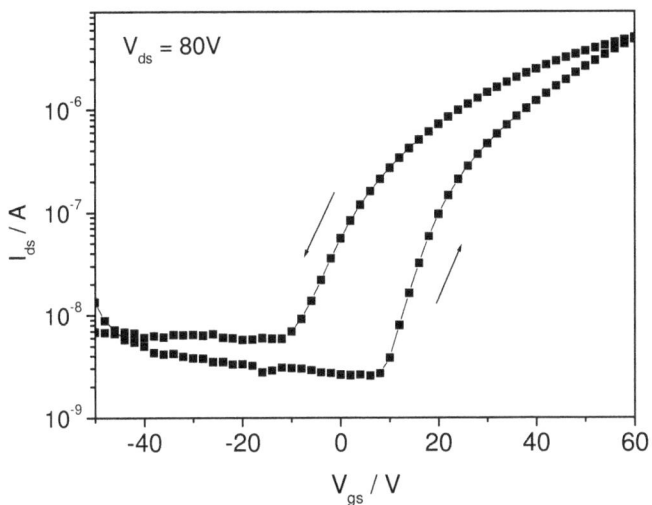

Fig. 4.12(a) Transfer characteristics of the PVA/MDMO-PPV: PCBM (1:4) blend based device in the dark at V_{ds} = 80 V. The arrows indicate the sweep direction, starting at – 50 V.

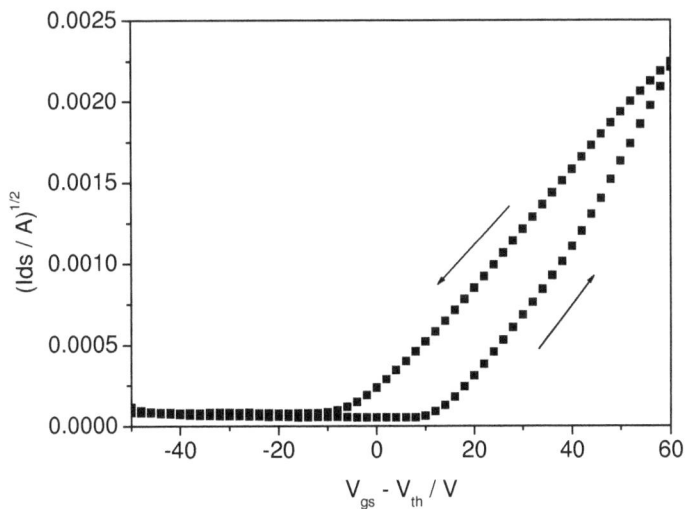

Fig. 4.12(b) $\sqrt{I_{ds}}$ vs. ($V_{gs} - V_{th}$) plot at V_{ds} = 80 V. The electron field effect mobility of 10^{-2} cm^2/Vs was calculated using Eq. (2) from the slope of the descending curve.

In addition to the hysteresis, in the PVA-photOFETs in dark a shift of the threshold voltage (bias-stress) towards higher voltages was observed, Fig. 4.13. Threshold voltage shifts are commonly attributed to a built-in electric field near to the dielectric/semiconductor interface caused by the presence of a sheet of charges [9, 10]. We assume that the same mechanism applies in our system.

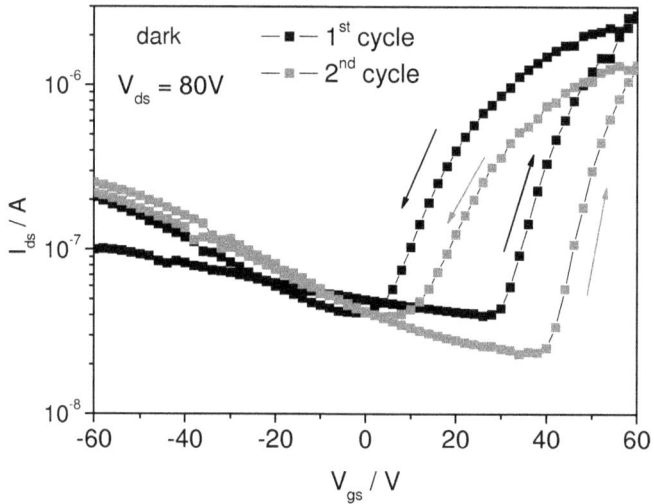

Fig. 4.13. Threshold voltage shift toward higher gate voltage in the PVA-photOFETs in the dark due to the gate-induced bias-stress. The arrows show the sweep direction, starting at – 60 V.

In an effort to observe ambipolar transport in this device the measurement of a hole enhanced current is performed by applying a negative V_{ds}, Fig. 4.14. As depicted, biasing the devices with a negative drain-source voltage (V_{ds} = -80 V) results in a non significant hole enhanced mode when LiF/Al source-drain electrodes are used.

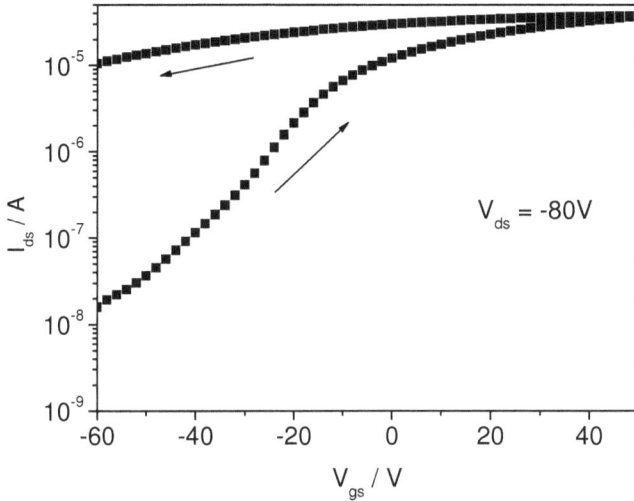

Fig. 4.14 Transfer characteristics of the PVA/MDMO-PPV: PCBM (1:4) in the dark at V_{ds} = -80 V. The current is plotted in absolute scale. The arrows show the sweep direction, starting at - 60 V.

4.3.2 Under monochromatic illumination

The transfer characteristics of the PVA-photOFETs in the dark and under monochromatic light (λ = 514 nm) illumination with different intensities are shown in Fig. 4.15. Transfer characteristic in the dark show an electron enhanced mode which is developed when the device is biased with positive voltages (curves with filled square symbols). Upon monochromatic illumination at a low light intensity (6 mW.cm^{-2}) the device shows transistor amplification. A shift in the threshold voltage towards larger positive voltages with respect to the dark value was observed. A similar behaviour under the same illumination conditions was also observed in the MIS devices based on PVA/MDMO-PPV: PCBM (1:4) blends (see Fig. 4.8). By increasing the illumination intensity up to 30 mW.cm^{-2}, the drain-source current, I_{ds}, becomes less gate dependent. By further increasing the illumination intensity up to 300 mW.cm^{-2}, I_{ds} becomes completely gate voltage independent, and a transition from a three to two terminal device occurs. At 1000 mW.cm^{-2} illumination intensity, the drain source current continues to be gate voltage independent. The device shows strong photodegradation at and after this high illumination intensity.

Transfer characteristics before and after illumination with a low light intensity are taken; Fig. 4.16 shows transfer characteristics of the initial dark state (curves with filled square symbols), under illumination of 6 mW.cm^{-2} (upper curve) and final state in the dark after illumination (curves with filled triangle symbols). As depicted, the threshold voltage shifts to larger positive bias voltage with respect to the initial dark state. Again, these results are in good correlation with the results obtained with similar MIS structures (see Fig. 4.8). the final dark state taken after exposing the device to the light reveals a significant and permanent device degradation.

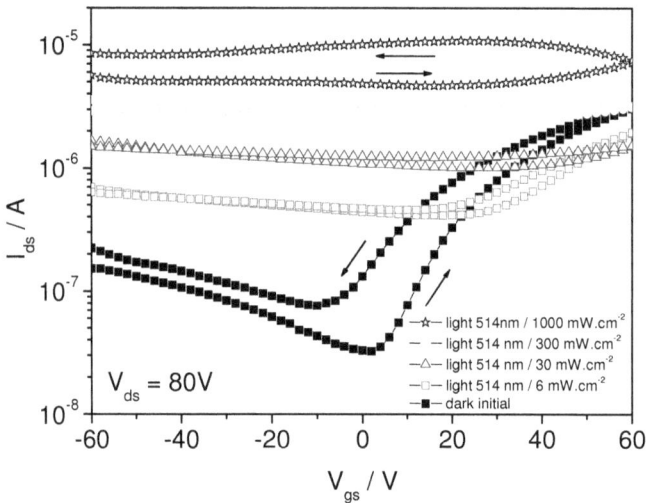

Fig. 4.15 Transfer characteristics of the PVA-photOFET in the dark (filled square curves) and under illumination (upper curves) taken at $V_{ds} = 80$ V. The arrows show the sweep direction, starting at − 60 V.

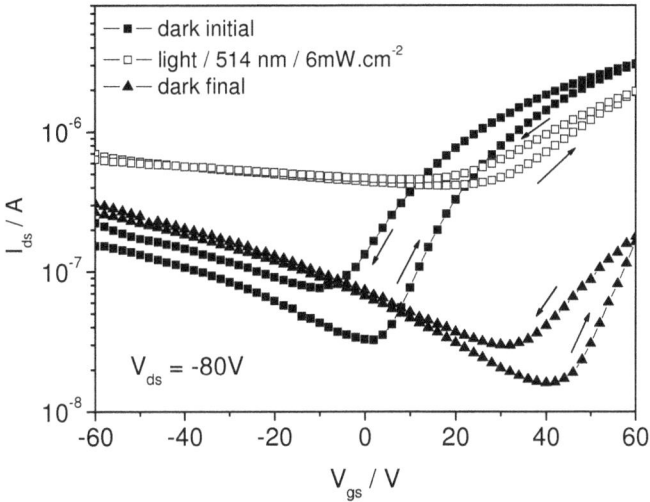

Fig. 4.16 Transfer characteristics of the PVA-photOFET in the dark initial state (filled square curves), under illumination with 6 mW.cm^{-2} (open symbol curves) and in the dark state after illumination (filled triangle symbol curves). The arrows show the sweep direction, starting at − 60 V.

4.3.3 Under AM1.5 illumination

The transfer characteristics of the PVA-photOFETs in the dark and under white light illumination are shown in Fig. 4.17 [11]. The curves with open square symbols (upper curve) in Fig. 4.17 show the *photoresponse* of the devices. At a low white light (with an AM1.5 wavelength spectrum) intensity of 1 mW/cm^2, the transfer characteristics show a gate voltage induced electron enhanced mode. In the depletion mode, I_{ds} increases more than two orders of magnitude in comparison to the dark currents. At higher light intensities the drain-source current, I_{ds} increases even more, becomes however gate voltage independent and the device performance changes to a two terminal photoresistor behaviour associated by a strong device degradation.

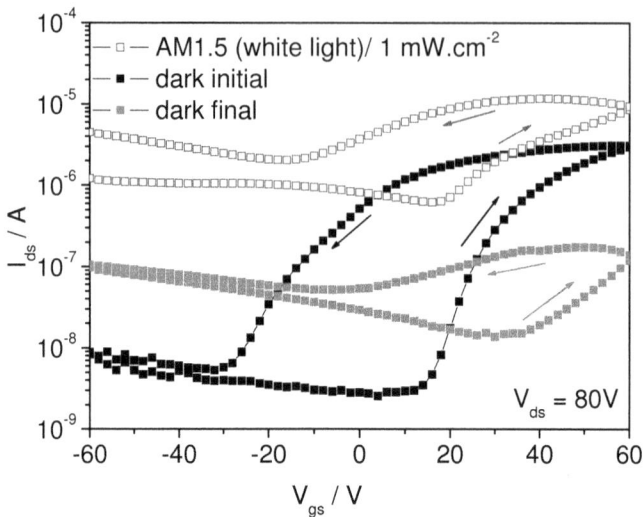

Fig. 4.17 Transfer characteristics of the PVA-photOFET in the dark (filled square curves), under AM1.5 (1 mW/cm^2) illumination (open symbol curves) and in the dark after illumination (filled triangle symbol curves) measured at $V_{ds} = + 80$ V. The arrows show the sweep direction, starting at – 60 V.

The increase in I_{ds} can be explained by the generation of a large number of charge carriers due to the photoinduced charge transfer at the conjugated polymer/fullerene bulk heterojunction (photodoping). In the illuminated PVA based photOFET devices, a high field is required for reaching the threshold voltage. After switching off the light, a significant shift in the transfer curve with respect to the initial dark transfer characteristic together with dramatic changes in the transfer characteristics was found (curves with filled triangular symbols in Fig. 4.17). The observed behaviour, seems to be a superposition of the light-induced bias-stress and the gate-induced bias-stress (Fig. 4.13), presumably due to complex interactions of different effects like charge trapping at the PVA/blend interface or in the bulk of the dielectric and semiconductor or from additional electric field induced charges. Further increase of illumination intensity result in dramatic difference in the dark initial and final state (hysteresis in transfer characteristic is minimized), Fig. 4.18. These results are well correlated with the results obtained on similar MIS structures (see Figs. 4.9 and 4.10). We attribute that behaviour to the photoinstability of the device and permanent device degradation. The origin and nature of the processes, which lead to the observed irreversibility, are not quite clear yet but we discussed some of our thoughts already in the last section.

50

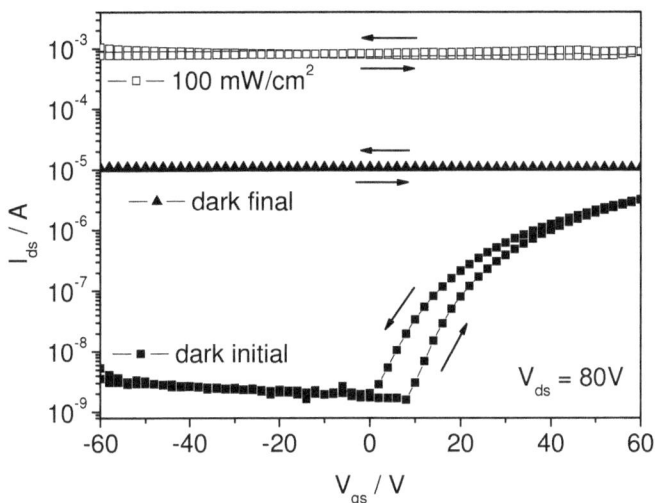

Fig. 4.18 Transfer characteristics of the PVA-photOFET in the dark state (curves with filled squares), under AM1.5 / 100 mW.cm^{-2} illumination (upper curve) and in the final dark state (curves with filled triangles). The arrows indicate the sweep direction, starting at – 60 V.

4.4 Summary

In summary, a series of devices fabricated on PVA were studied. Among polymeric gate dielectrics, PVA shows a high dielectric constant and forms highly transparent films. The depositing process is simple and easy to perform under ambient environment conditions at room temperature. Therefore, PVA is a promising candidate dielectric to be used as gate insulator in OFETs used in the dark.

The observed frequency dependent capacitance in pristine PVA based MIM devices shows fingerprints of charged species present in the bulk of the dielectric.

MIS structures based on MDMO-PPV: PCBM (1:4) blends and PVA were characterized in the dark, under monochromatic and under white light illumination. Clear electron injection and accumulation was observed in the devices. A significant hysteresis in *C-V* characteristics in the dark and under illumination was observed, presumably due to charge trapping and detrapping at the PVA/blend interface. Upon illumination, photogenerated charge carriers cause an increase in the sample capacitance. A shift of the threshold voltage towards higher positive voltage upon illumination was explained by

.

requiring a larger field for charge detrapping. Under white light illumination, photoinstability and permanent device degradation during and after measurement was observed.

PhotOFETs based on PVA/MDMO-PPV: PCBM (1:4) blends are fabricated and characterized under similar conditions as the MIS devices. In the dark, transistors with LiF/Al top source-drain contacts showed n-type behavior, with a significant hysteresis in the transfer characteristics and the threshold voltage shift (bias-stress). Under monochromatic or white light illumination; PVA-photOFETs shows a relatively high photoresponse, but a weak photostability.

Nevertheless, the observed hysteresis in PVA has been utilized in memory elements in the dark (memOFETs) as reported in Chapter 6.

4.5 References

[1] B.C. Shekar, V. Veeravazhunthi, S. Sakthivel, D. Mangalaraj, and Sa.K. Narayandass, *Thin Solid Films*, 384, 122, 1999.

[2] X. Yang, J. K. J. van Duren, R. A. J. Janssen, M. A. J Michels, and J. Loos, *Macromolecules* 37, 2151, 2004.

[3] H. Hoppe, M. Niggemann, C. Winder, J. Kraut, R. Hiesgen, A. Hinsch, D. Meissner, and N. S. Sariciftci, *Adv. Funct. Mater.* 14, 1005, 2004.

[4] Th. B. Singh, N. Marjanović, P. Stadler, M. Auinger, G. J. Matt, S. Günes, N. S. Sariciftci, R. Schwödiauer, and S. Bauer, *J. Appl. Phys.* 97, 083714, 2005.

[5] Th. B. Singh, N. Marjanović, G. J. Matt, N. S. Sariciftci, R. Schwödiauer, and S. Bauer, *Appl. Phys. Lett*, 85, 5409, 2004.

[6] L.L. Chua, J. Zaumsell, J.-F. Chang, E. C.-W. Ou, P. K.-H. Ho, H. Sirringhaus, and R. H. Friend, *Nature*, 434, 194, 2005.

[7] V. D. Mihailetchi, J. K. J. van Duren, P. W. M. Blom, J. C. Hummelen, R. A. J. Janssen, J. M. Kroon, M. T. Rispens, W. J. H. Verhees, and M. M. Wienk, *Adv. Func. Mater.* 13, 43, 2003.

[8] G. J. Matt, N. S. Sariciftci, and T. Fromherz, Appl. Phys. Lett. 84, 1570 (2004); C. J. Brabec, A. Cravino, D. Meissner, N. S. Sariciftci, T. Fromherz, M. T. Rispens, L. Sanchez, and J. C. Hummelen, *Adv. Funct. Mater.* 11, 374, 2001.

[9] S.M. Sze, Physics of Semiconductor Devices, Wiley-Interscience, New York, 1981.

[10] A. Salleo and R.A. Street, *J. Appl. Phys.* 94, 471, 2003.

[11] N. Marjanović, Th. B. Singh, G. Dennler, S. Günes, H. Neugebauer, N. S. Sariciftci, R. Schwödiauer, and S. Bauer, *Org. Elect.* in press

Chapter 5

5. photOFETs based on MDMO-PPV: PCBM (1:4) blends on top of BCB gate-insulator

PhotOFETs based on BCB and MDMO-PPV: PCBM (1:4) blends, BCB-photOFETs, will be presented in this Chapter. As top source and drain contacts, LiF/Al were used. The BCB/blend interface was studied with *C-V* measurements on MIM and MIS devices. Again, prior to the transistors results observations on MIM and MIS devices will be discussed.

5.1. BCB based MIM device

Metal-Insulator-Metal (MIM) devices based on pristine BCB dielectrics were fabricated and characterised as described in Chapter 2.

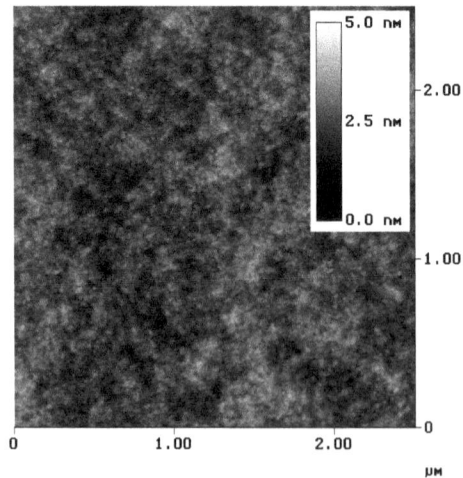

Fig. 5.1 AFM image of the BCB dielectric film.

Fig. 5.1 shows the height image of spin coated and crosslinked BCB dielectric films obtained by AFM measurements in tapping mode. It can be concluded from the figure that BCB provides a smooth surface with a roughness below 5 nm.

The capacitance vs. frequency of the BCB based MIM device in the dark is shown in Fig. 5.2. As shown, no frequency dependence of the capacitance was observed. BCB was cured after spin coating in order to create the network polymer structure. In comparison to PVA, the amount of impurities in the crosslinked BCB is smaller. This is one of the reasons for the excellent insulating, dielectric, mechanical, thermal and chemical properties of BCB. A dielectric constant of 2.65 was estimated from BCB based MIM devices.

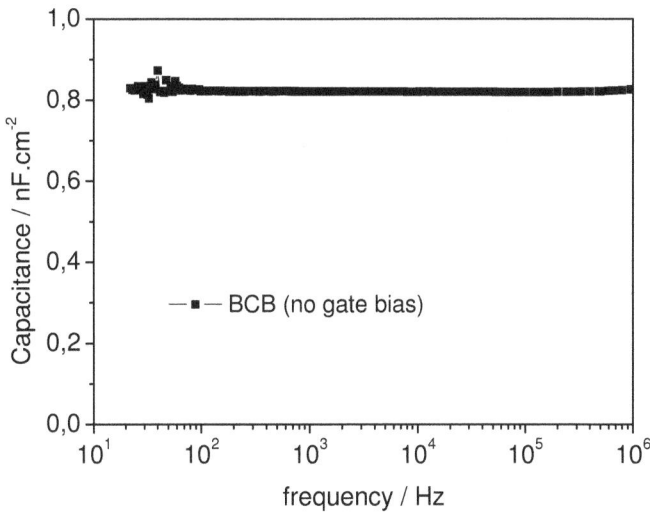

Fig. 5.2 Capacitance vs. frequency of BCB based MIM devices.

5.2. BCB/MDMO-PPV: PCBM (1:4) blend based MIS devices

5.2.1. Dark conditions

Metal-Insulator-Semiconductor (MIS) devices based on BCB and MDMO-PPV: PCBM (1:4) blends were fabricated and characterised in the dark and under illumination.

The height image of spin coated MDMO-PPV: PCBM (1:4) blends on BCB dielectric films obtained by AFM investigations in tapping mode is shown in Fig. 5.3. Phase separation in the blend film similar to the blend film on top of PVA is also observed.

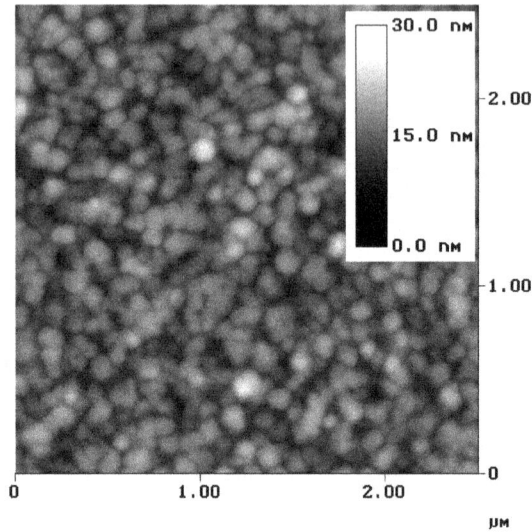

Fig. 5.3 AFM image of the BCB/MDMO-PPV: PCBM (1:4) blend film.

The capacitance vs. frequency of the BCB/MDMO-PPV: PCBM (1:4) blend based MIS device is shown in Fig. 5.4. With positive bias voltage, injected electrons, accumulated at the BCB/blend interface, induce an increase of the capacitance. In contrast, negatively biasing the device shows no significant change in the capacitance, similar to the case of PVA-MIS devices.

Fig. 5.5 shows C-V characteristics of the BCB/MDMO-PPV: PCBM (1:4) based MIS device in the dark. The electron injection and accumulation starts around 0 V: By positively biasing the device an increase in the capacitance occurs. A negligible hysteresis is observed when changing the sweep direction. By increasing the frequency range, the capacitance shows a decreasing trend due to the fact that the charge carriers cannot follow the fast ac-voltage modulation.

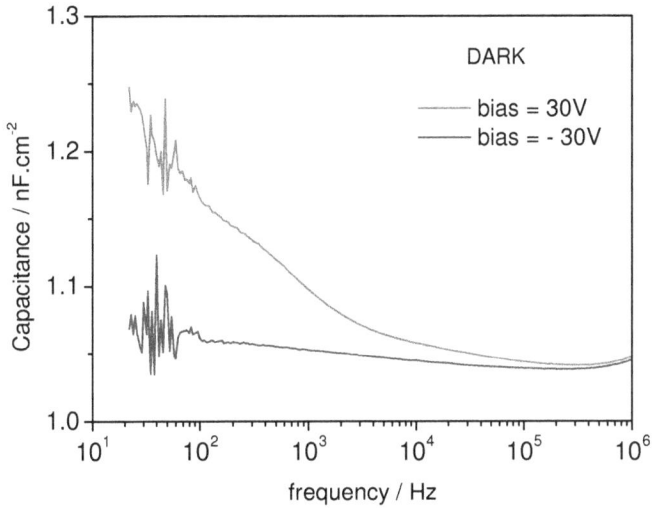

Fig. 5.4 Capacitance vs. frequency of BCB/MDMO-PPV: PCBM (1:4) based MIS devices.

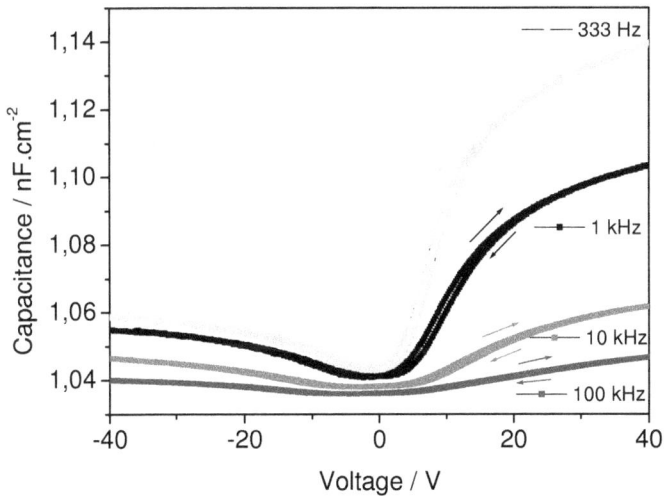

Fig. 5.5 *C-V* characteristics of BCB/MDMO-PPV: PCBM (1:4) based MIS devices in the dark. The arrows show the sweep direction, starting at – 40 V.

5.2.2. Under monochromatic illumination

The MIS devices were also characterised under monochromatic illumination conditions. As a monochromatic light source, an Ar^+ laser was used with a wavelength of 514 nm and an intensity 6 mW.cm^{-2}. The transient capacitance measurements for fixed light intensity are shown in Fig. 5.6. The device was biased with 0 V and +20 V, corresponding to the accumulation-operating mode of the device. Again, the light to dark ratio in the accumulation regime is smaller than in the depletion regime since the electrons injected by the contact are contributing to the total capacitance. This kind of behaviour was already observed in PVA based MIS devices. The observed device behaviour can be used for tuning the device photoresponsivity of the device simply by adjusting the bias voltage.

In Fig. 5.7, a transient capacitance characteristic is shown. A measurable difference in the capacitance before and after illumination is observed in the accumulation mode, presumably due to charge trapping. A similar effect is observed in blend based MSM diodes (Fig. 3.6). It will be shown in the case of transistors that those charges can be de-trapped either by annealing or by applying large negative bias voltages. In the case of MIS devices, de-trapping was observed by leaving the device over night.

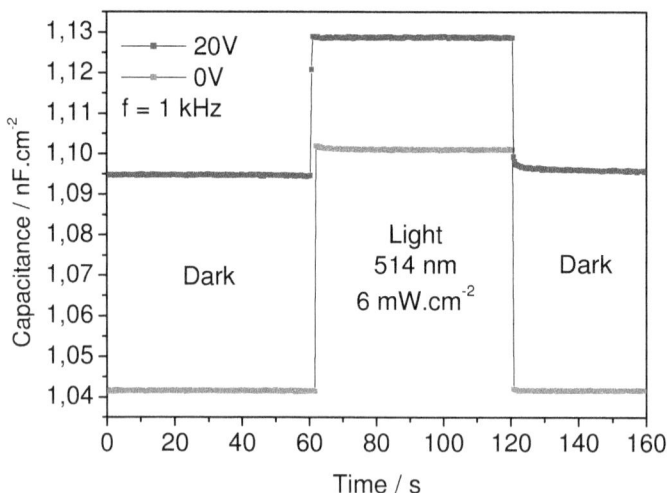

Fig. 5.6 Transient capacitance of BCB/MDMO-PPV: PCBM (1:4) based MIS devices.

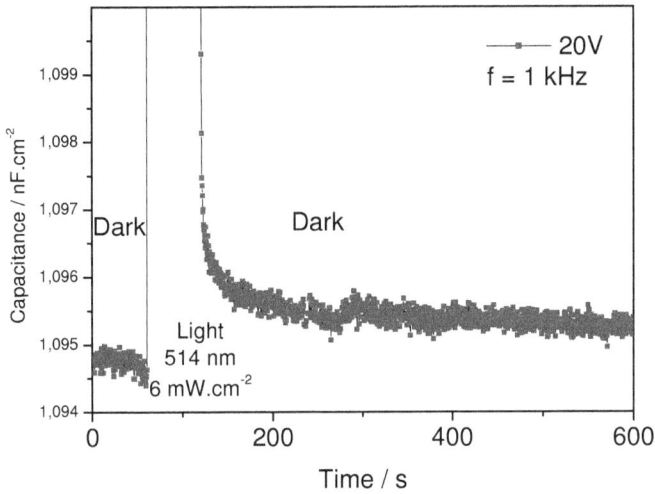

Fig. 5.7 Transient capacitance of the devices at +20 V.

C-V characteristics taken under monochromatic illumination are shown in Fig. 5.8. An increase in the capacitance due to photodoping and a negligible hysteresis is observed.

Fig. 5.9 shows initial dark, illumination and final dark *C-V* characteristics. A clear increase in the capacitance upon illumination due to the contribution of photoinduced charge carriers to the total capacitance is observed. Again, a shift of the threshold voltage caused by trapping of the photogenerated charges is seen, but is much smaller in comparision to PVA based MIS devices (Fig. 4.8).

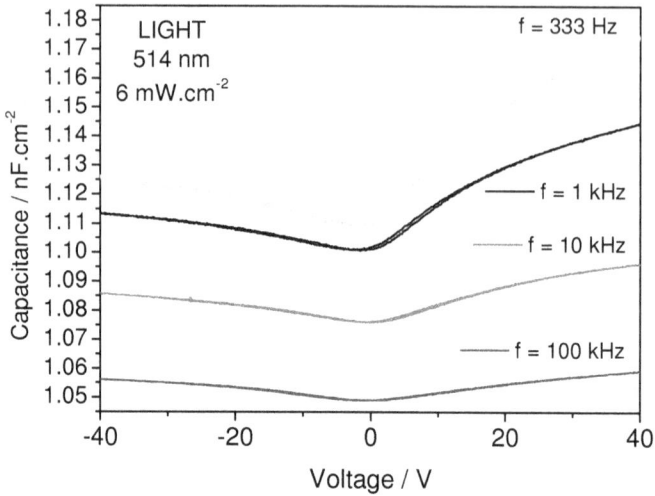

Fig. 5.8 *C-V* characteristic of BCB/MDMO-PPV: PCBM (1:4) blend based MIS devices under monochromatic illumination.

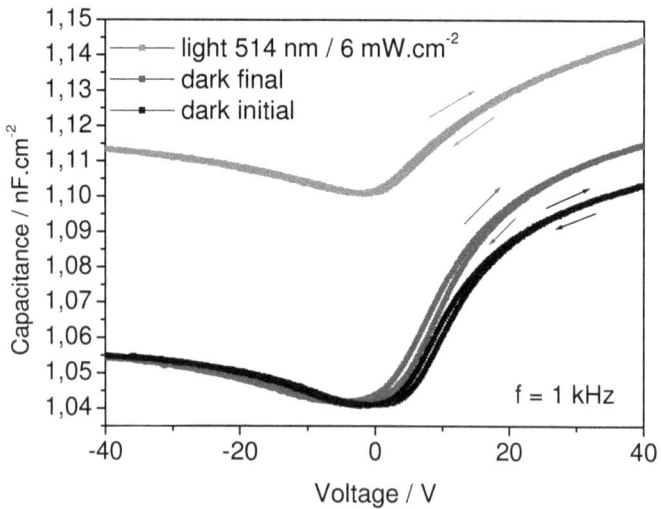

Fig. 5.9 *C-V* characteristics of the BCB/MDMO-PPV: PCBM (1:4) blend based MIS devices in the dark, under monochromatic illumination and in the dark after illumination at 1 kHz. The arrows show the sweep direction, starting at – 40 V.

5.2.3. Under AM1.5 illumination

Fig. 5.10 shows *C-V* characteristics of the MIS device in the dark (dark initial), under white light (AM1.5) illumination, and in the dark after illumination (dark final).

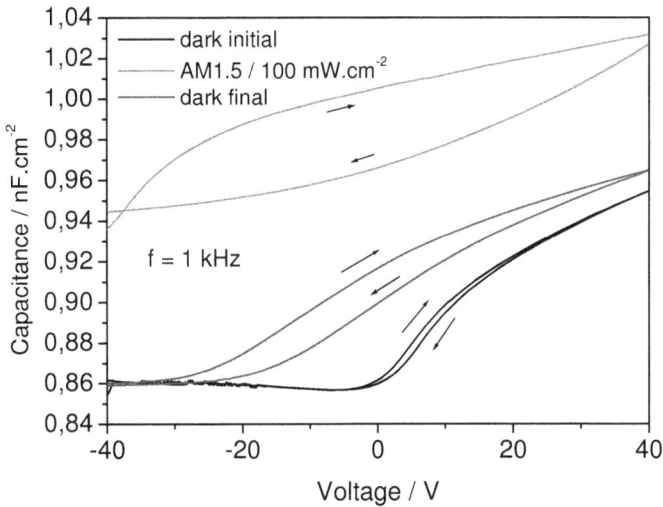

Fig. 5.10 *C-V* characteristics of the MIS device in the dark, under white light and in the dark after white illumination. The arrows show the sweep direction, starting at – 40 V.

In general, the device response is similar to the device response under monochromatic illumination. The increase of the capacitance caused by photogenerated space charges is observed together with a shift of the threshold voltage in comparison to the initial state. The threshold voltage shifts towards more negative values, as compared to the shift under monochromatic illumination. Also, the increase in the capacitance due to illumination is higher here. Again, there is no sign of hole injection. The hysteresis in the *C-V* characteristics under white light illumination becomes more pronounced. Also, we cannot exclude heating effects from the solar simulator itself, which may change the charge trapping/de-trapping kinetics.

5.3. BCB/MDMO-PPV: PCBM (1:4) blend based photOFETs

5.3.1. Dark conditions

Output characteristics of the BCB - MDMO-PPV: PCBM (1:4) blends based photOFETs, with LiF/Al as top source and drain electrodes in the dark are shown in Fig. 5.11. As mentioned above, crosslinked BCB forms an inert dielectric layer with excellent dielectric properties. An electron enhanced mode in the dark is achieved by biasing the devices with positive gate-source voltages, V_{gs}.

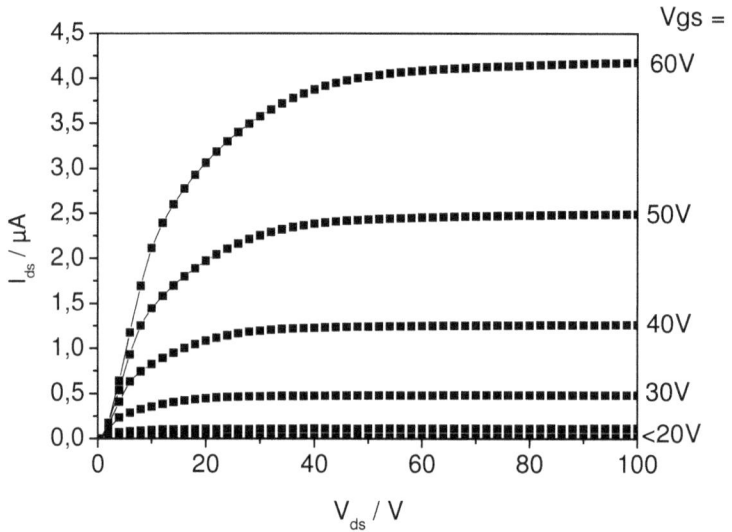

Fig. 5.11 Output characteristics of the MDMO-PPV: PCBM (1:4) based photOFET fabricated on top of the BCB in the dark.

Transfer characteristics of the same device in the dark show n-type unipolar transistor behaviour, Figs. 5.12(a),(b) and 5.14. A negligible hysteresis in the transfer curves at $V_{ds} = 80$ V is observed, Fig. 5.12(a). From the slope of the $\sqrt{I_{ds}}$ vs. (V_{gs}- V_{th}) plot Fig. 5.12(b), an electron field-effect mobilitiy, μ_e of 10^{-2} cm^2/Vs was calculated by using Eq. (2). No gate-induced threshold voltage shift [10] was observed in the BCB-photOFETs in the dark, Fig. 5.13. As in the case of PVA/blend based OFETs, in order to check for ambipolar transport in BCB/blend based OFETs, measurements of the hole enhanced current was performed. By applying a negative drain-source voltage, no hole accumulation is observed in the device, as shown in Fig. 5.14.

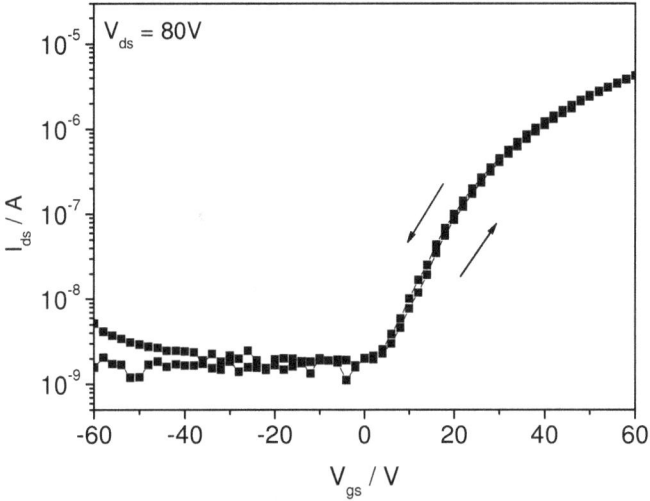

Fig. 5.12(a) Transfer characteristics of the BCB/MDMO-PPV: PCBM (1:4) blend based photOFET in the dark at $V_{ds} = 80$ V. The arrows show the sweep direction, starting at -60 V.

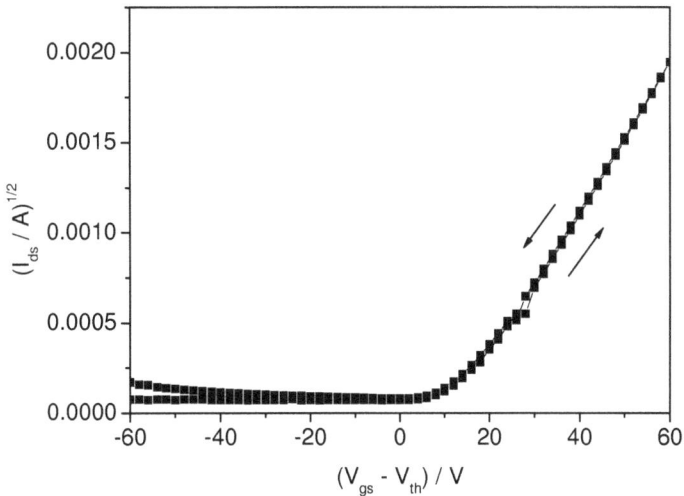

Fig. 5.12(b) $\sqrt{I_{ds}}$ vs. $(V_{gs} - V_{th})$ plot at $V_{ds} = 80$ V. The electron field effect mobility of 10^{-2} cm^2/Vs was calculated from the slope of the curve by using Eq. (2)

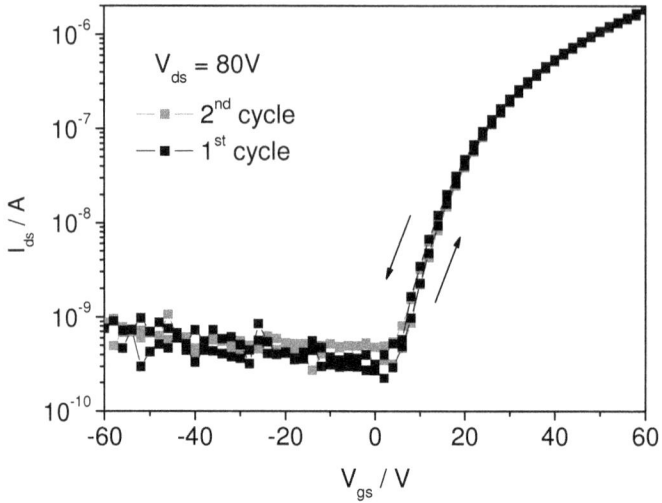

Fig. 5.13 Two cycles of the transfer characteristics of the BCB/MDMO-PPV: PCBM (1:4) blend based photOFETs in the dark shows no gate-induced threshold voltage shift. The arrows show the sweep direction, starting at – 60 V.

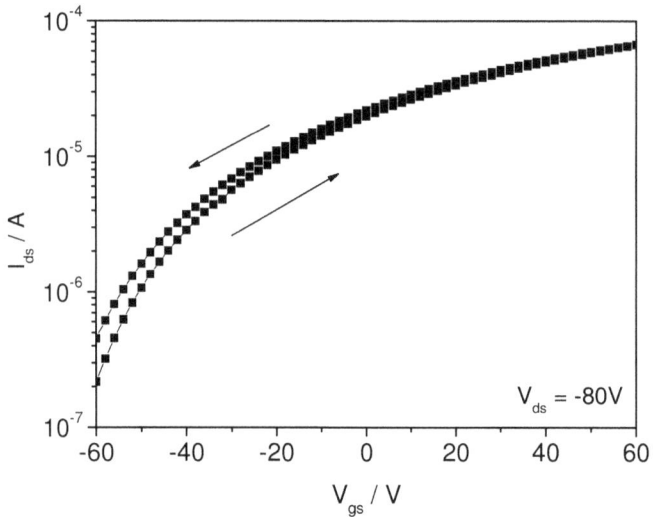

Fig. 5.14 Transfer characteristics of the BCB/MDMO-PPV: PCBM (1:4) blend based photOFET in the dark at V_{ds} = -80 V. The absolute value of the current is plotted vs. V_{gs}. The arrows show the sweep direction, starting at – 60 V.

5.3.2. Under monochromatic illumination

The light response of the BCB-photOFET is shown in Fig. 5.15 by comparing the drain-source current obtained in the dark (curves with filled symbols) and under given monochromatic illumination (curves with open symbols) for different gate voltages. Upon illumination, photoinduced charge transfer between the conjugated polymer and the fullerene in the blend occurs, increasing the number of the charge carriers which cause increase in I_{ds}. The negligible hysteresis in the transfer characteristics in the dark and under monochromatic illumination is presented in Fig. 5.16.

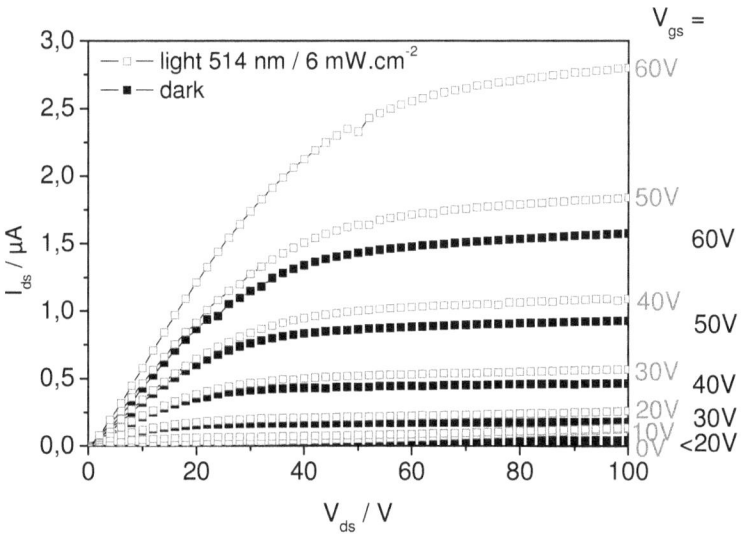

Fig. 5.15 Output characteristics of the BCB/MDMO-PPV: PCBM (1:4) based photOFET in the dark (curves with filled squares) and under monochromatic illumination (curves with open squares) for different gate voltages.

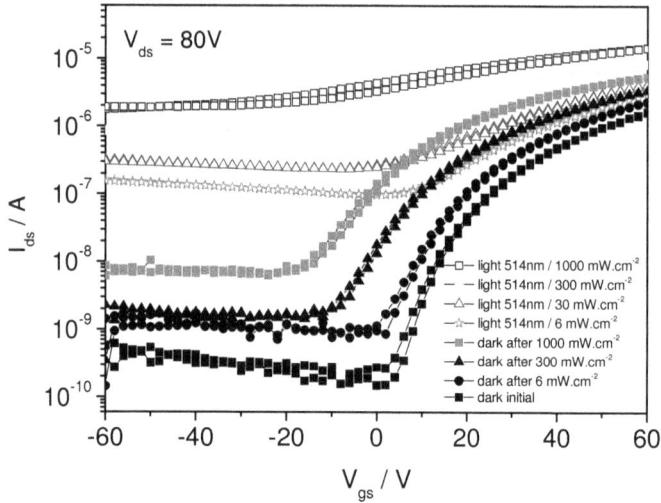

Fig. 5.16 Transfer characteristics of the above shown device under monochromatic light illumination with different light intensities at $V_{ds} = 80$ V.

By increasing the illumination intensity from 6 mW.cm^{-2} to 1000 mW.cm^{-2}, a shift in the threshold voltage is observed, presumably due to light-induced bias-stress [10]. No gate induced bias-stress was observed in the BCB-photOFETs in dark as shown previously (Fig. 5.13). At 1000 mW.cm^{-2} of focused laser light the device still shows transistor amplification behaviour. The dark transfer characteristic after exposure of the device to that illumination intensity was taken (curves with red filled squares, Fig. 5.16) demonstrating that the device is still operating. The off current becomes higher with respect to the initial dark curves. A shift of the threshold voltage towards negative gate bias voltages and an increase in the off current is observed, presumably due to the light induced charge trapping at the BCB/blend interface (light-induced bias-stress). Clearly observable is that in contrast to PVA, BCB is a gate-insulator forming a much more photostable system.

5.3.3. Under AM1.5 illumination

As mentioned above, crosslinked BCB forms an inert dielectric layer with excellent mechanical properties and chemical stability. On top of BCB bulk heterojunction MDMO-PPV: PCBM (1:4) blends based photOFETs were fabricated. As top source and drain electrodes, LiF/Al was used. An electron enhanced mode in the dark and under AM1.5 (100

66

mW/cm^2) illumination is achieved by biasing the devices with positive gate-source voltages, V_{gs} (Fig. 5.17) [1]. The light response of the devices is clearly revealed by comparing the values of the drain-source current in the dark and under illumination. Again, the increase of I_{ds} is caused by the creation of a large number of free charge carriers due to photoinduced charge transfer between the conjugated polymer and the fullerene in the blend. A negligible hysteresis in the initial dark transfer characteristics is observed (curves with filled squares in Fig. 5.18) [1]. A calculated electron mobility, μ_e of 10^{-2} cm^2/Vs is derived from the initial dark transfer characteristics at V_{ds} = +80 V. The transfer characteristics of the device upon illumination with white light (AM1.5) and under different illumination intensities (from 0.1 – 100 mW/cm^2) are shown in Fig 5.18 with open symbols. The drain-source current in the depletion regime of the device is significantly increased upon illumination. In the accumulation regime, the increase of the drain-source current is less pronounced. The photOFET *responsivity R*, calculated from equation (6) (Chapter 1) in the depletion region was found to be 10 mA/W and in the accumulation regime 0.15 A/W, respectively. The *photosensitivity P* and *photoresponse $R_{I/d}$*, (equations 7 and 8, respectively, Chapter 1) have a maximum in the strong depletion regime, 10^2 and a minimum in the strong accumulation regime, 10^0.

The threshold voltage for reaching the accumulation mode and for opening the transistor shifts to lower values upon illumination, suggesting that the trap carrier density in the channel is enhanced by photodoping. The higher responsivity in the accumulation regime than in the depletion regime is attributed to the number of photogenerated charge carriers in the blend, which depends mostly on the light intensity and not on the applied gate voltage.

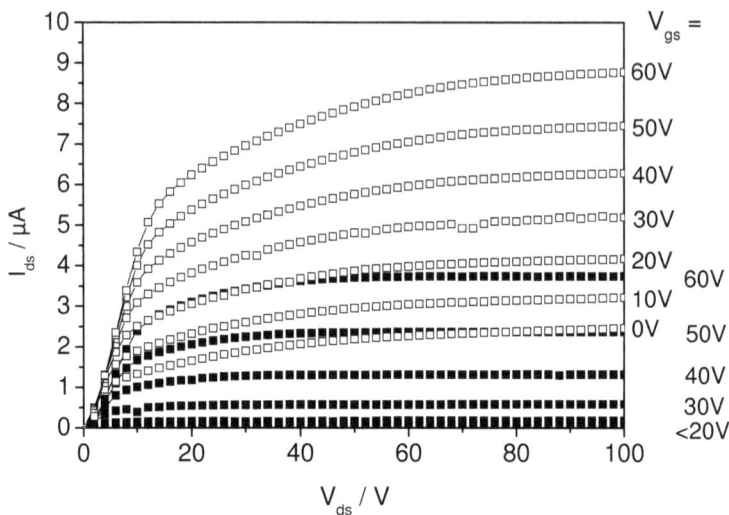

Fig. 5.17 Output characteristics of the MDMO-PPV: PCBM (1:4) based photOFETs in the dark (filled squares) and under AM1.5 (100 mW/cm^2) (open squares).

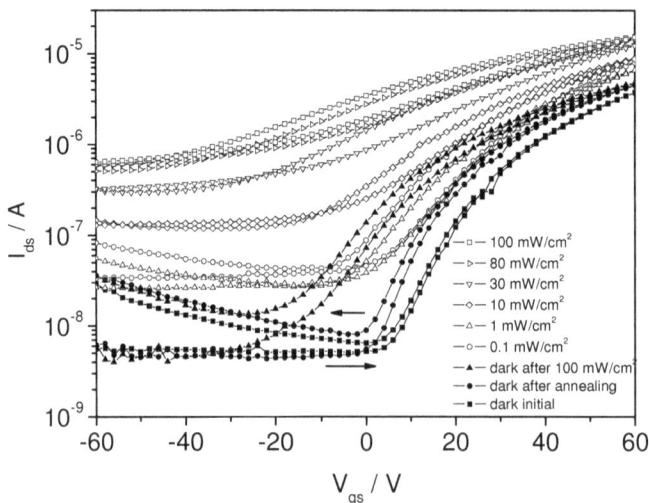

Fig. 5.18 Transfer characteristics of the same device in the dark (filled squares), under AM1.5 illumination for different illumination intensities from 0.1 to 100 mW/cm^2 (open symbols), in the dark after illumination (filled triangles) and in the dark after annealing at 130^0C for 3 min (filled circles), measured at $V_{ds} = +80$ V.

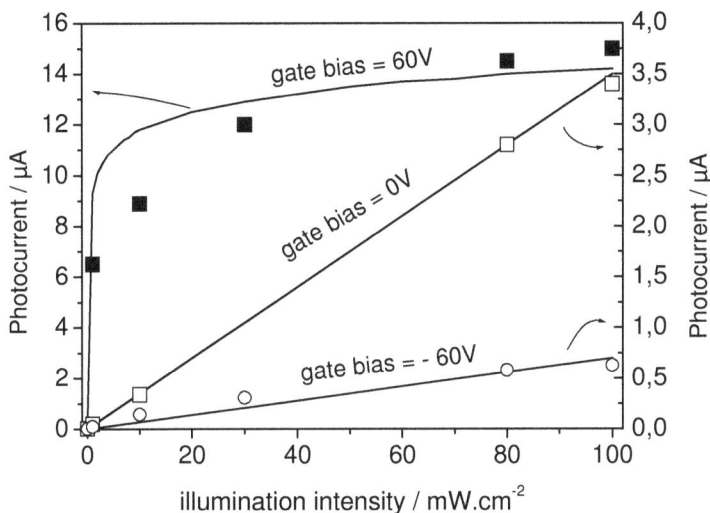

Fig. 5.19 Photocurrent as a function of the illumination intensity in the ON-state (V_{gs} = 60 V, filled square symbols), under low gate bias (V_{gs} = 0 V, open square symbols) and in the OFF-state (V_{gs} = -60 V, open circle symbols). Solid lines are the fits based on Eqs. (4) and (5) (Chapter 1)

Upon illumination two different effects may occur in the active layer of the transistor, *photoconductivity* and the *photovoltaic effect*. When the transistor is in the ON-state the photocurrent is dominated by the photovoltaic effect and is given by Equation (4) (Chapter 1). The photovoltaic effect together with a shift of the threshold voltage is caused by the large number of trapped charge carriers located under the source electrode and within the semiconductor near the dielectric interface [2-5, 10].

When the transistor is in the OFF-state, the photocurrent induced by photoconductivity is given by Equation (5) (Chapter 1). The experimental result of photOFETs based on MDMO-PPV: PCBM (1:4) blends and BCB (Fig. 5.19) are in good agreement with the calculations based on Eqs. (4) and (5) [1].

After illumination, a shift in the dark transfer curve with respect to the initial dark transfer characteristics was observed (curves with filled triangular symbols Fig. 5.18) presumably due to persistent interface effects, *i.e.* charge trapping, bias stressing, etc. [6-10]. In contrast to PVA, the initial dark state was recovered (*e.g.* recovery if the device by removing the trapped charges) by annealing the device at 130°C for 3 minutes (curve with filled circles, Fig.5.18). Fig. 5.20 shows the drain-source current recorded sequentially at V_{gs}

= 0 V in the dark, under AM1.5 (100 mW/cm²) illumination, again in the dark and after annealing, for the first three cycles [1]. Similar values for the respective dark/illumination cycles show the reversibility of the device. The existence of a state in the device with a larger current after illumination and prior to annealing may be considered as a memory effect, which can be set by light and erased either by annealing or by applying large negative gate voltages. Light memory induced effect in the photOFETs will be discussed in Chapter 6.

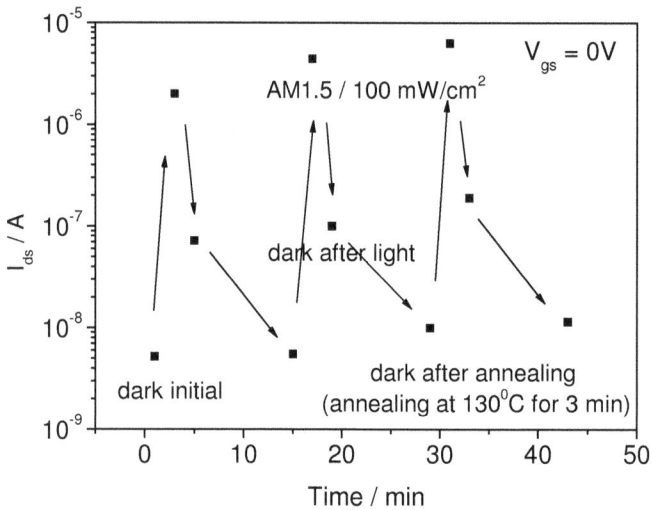

Fig. 5.20 Drain-source current recorded at $V_{gs} = 0$ V in the dark, under AM1.5 (100 mW/cm²) illumination, again in the dark, and after annealing, for the photOFETs presented in Fig.5.17.

Finally, transfer characteristics in the negative drain-source bias regime ($V_{ds} = -80$ V) of the BCB-photOFETs in the dark and under white light illumination are shown in Fig. 5.21. As shown before, there is no hole accumulation regime achieved neither in the dark nor under illumination.

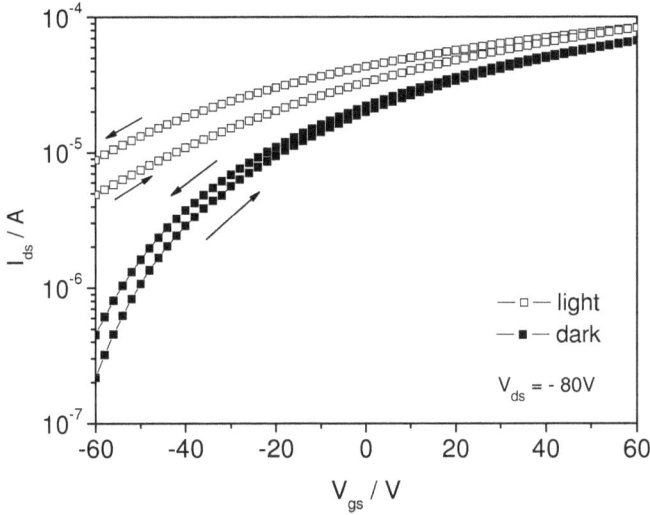

Fig. 5.21 Transfer characteristics of the BCB-photOFET in the dark and under white light illumination at V_{ds} = -80 V. The arrows show the sweep direction, starting at – 60 V.

5.4. Spectral characterisation of the BCB/MDMO-PPV: PCBM (1:4) blend based photOFETs

Recent reports of the optoelectrical characterisation of phototransistors based on poly(3-hexythiophene) (P3HT) [11] and on pentacene [12,13] show that spectroscopy may find general applications in the analysis of the performance of organic FETs.

The detailed experimental part related to the spectral characterisation of the BCB-photOFET is given in Chapter 2. The spectral dependency of the photogenerated current in the BCB-photOFET together with the absorption spectra of the MDMO-PPV: PCBM (1:4) blend (dashed line) is shown in Fig. 5.22. The spectrum in Fig. 5.22 was obtained by recording the transistor transfer characteristics at selected wavelengths and making a cross section at a certain V_{gs} (constant V_{ds} = 80 V). The dark current was subtracted from these curves to show only the light induced current, which is normalized to the incident light power kept constant for all wavelengths. The constant light power is necessary to compare the current values at different wavelengths, as the transistor response is nonlinearly dependent on the incident light power (see § 5.3.3., Chapter 5).

The absorption spectra of the blend shows two maxima at ~ 360 nm and ~ 480 nm, which can be identified as absorption peaks originating from PCBM and MDMO-PPV, respectively. In general, $I_{ph}(\lambda)$ at $V_{gs} = 0$ V follows the absorption spectrum. In the accumulation regime, a strong amplification of the photogenerated current due to the applied gate voltage is observed, whereas in the depletion regime, it shows no gate voltage dependence.

Furthermore, in the strong depletion regime, at $V_{gs} = -40$ V and -60 V, the contribution of PCBM to the photogenerated current seems to be suppressed. The noisy current in the spectral region of low absorption (600 – 800 nm) is probably due to stray light and the small, but significant absorption of PCBM in this spectral range.

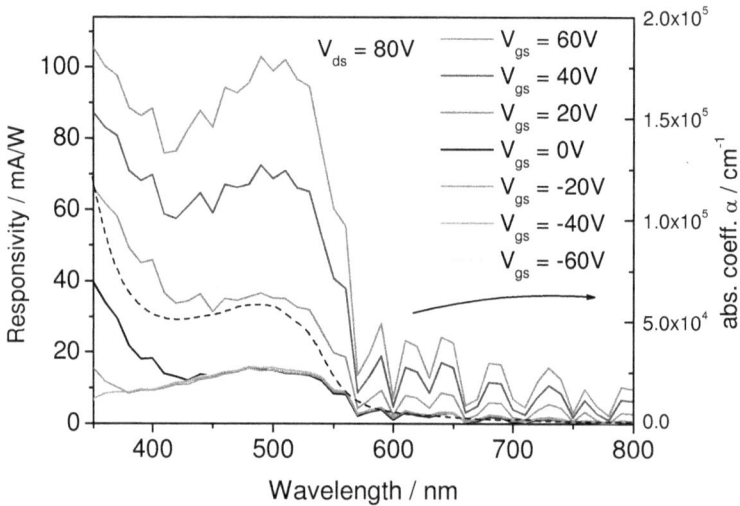

Fig. 5.22 Spectral response of a BCB-photOFET; absorption coefficient of the MDMO-PPV: PCBM (1:4) blend (dashed line).

The photosensitivity (given by equation 7, Chapter 1) is higher in the depletion than in the accumulation regime, as already mentioned in § 5.3.3. This is due to the fact that the amplification of the dark drain-source current in the accumulation regime is higher than the amplification of the light-induced current. The highest photosensitivity is reached without any gate voltage, a state corresponding to a two-terminal photodiode. The photosensitivity versus gate voltage for a wavelength of 490 nm is given in Fig. 5.23.

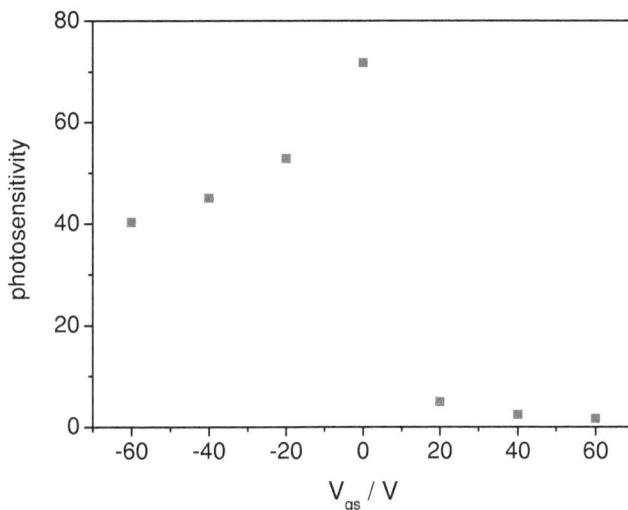

Fig. 5.23 Photosensitivity vs. gate voltage of a BCB-photOFET at $\lambda = 490$ nm.

The transfer characteristics in the dark were recorded sequentially after illumination, as depicted in Fig. 5.24. The initial dark transfer characteristics are taken prior to the illumination of the device (dark start before 800 nm in Fig. 5.24). An important point is that a threshold voltage shift towards negative voltages occurs after illumination with 410 nm. An even more pronounced threshold voltage shift and a hysteresis in the transfer characteristics (light-induced bias-stress) occurs after illumination of the device with 350 nm. The threshold voltage shift and appearance of a hysteresis was already observed in the *C-V* characteristics of the BCB/blend based MIS device (Fig. 5.10) and in the transfer characteristics of the BCB-photOFETs taken during and after AM1.5 illumination (Fig. 5.18). Present results indicate that charges originating from a photoexcitation in the blue spectral region are causing a shift of the threshold voltage (light-induced bias-stress) and the hysteresis in the BCB/MDMO-PPV: PCBM systems. This seems to be connected with the suppression of the UV contribution in the photocurrent spectra, Fig. 5.22.

The light memory element will be discussed in Chapter 6.

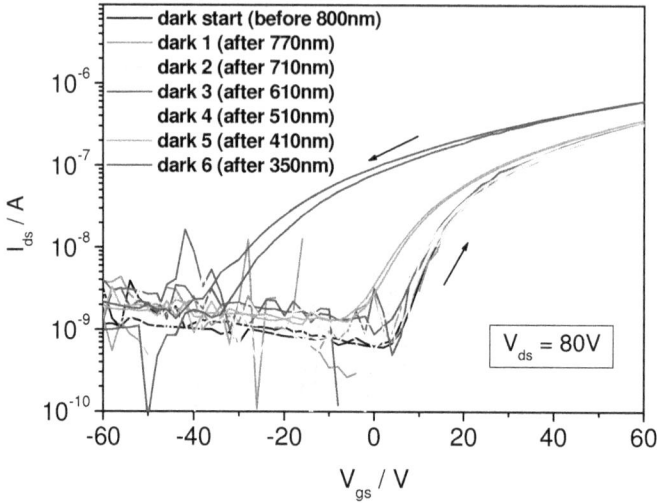

Fig. 5.24 Transfer characteristics of a BCB-photOFET at $V_{ds} = 80$ V in the dark taken sequentially during the optoelectrical measurement. The arrows show the sweep direction, starting at -60 V.

5.5 Summary

Crosslinked BCB forms an inert dielectric layer with excellent mechanical properties and high transparency. Series of devices fabricated on BCB gate insulator were studied.

In contrast to PVA, no frequency dependence in the capacitance in pristine BCB based MIM device was observed.

MIS structures based on MDMO-PPV: PCBM (1:4) blends on BCB were characterized in the dark, under monochromatic and under white light illumination. Clear electron injection and accumulation was observed in those devices. A negligible hysteresis in the C-V characteristics was observed in dark. Upon illumination with monochromatic or white light, photogenerated charge carriers cause capacitance increases in MIS devices. Trapping of photogenerated charge carrier explained the shift of the threshold voltage towards negative voltages upon illumination. Relatively smaller increases of the capacitance in the accumulation regime than in the depletion regime upon illumination can be used in practical application for tuning the device photoresponsivity by adjusting the bias voltage. Under white light illumination, increases of the capacitance were higher than in the case of monochromatic illumination. A hysteresis in the *C-V* characteristics under white light illumination appears, presumably because of charge trapping.

PhotOFETs based on BCB/MDMO-PPV: PCBM (1:4) blends are fabricated and characterized under similar conditions as the MIS devices. In the dark, transistors with LiF/Al top source-drain contacts showed n-type behavior, with negligible hysteresis in the transfer characteristics and no gate-induced threshold voltage shift. Under monochromatic illumination; BCB-photOFETs shows high response and relatively good photostability, even at very high illumination intensities (up to 1000 mW.cm^{-2}). Under white light illumination, an increase in I_{ds} of more that two orders of magnitude in the depletion regime caused by the generation of a large number of charge carriers is observed. A photovoltaic effect together with a shift of the threshold voltage, caused by the large number of trapped charge carriers under the source electrode and near to the dielectric/semiconductor interface (light-induced bias-stress), in the transistor ON-state, is observed. When the transistor is in the OFF-state, the photocurrent is induced by a photoconductivity. The obtained results show good correlation with the theory.

After illumination of the BCB-photOFETs, a shift of the dark transfer curve with respect to the initial dark transfer curve was observed, presumably due to light-induced bias-stress occur at the BCB/MDMO-PPV: PCBM interface. Recovery of the initial dark state (*e.g.* recovery of the device by removing the trapped charges) was achieved by annealing. It is proposed to exploit the effect of an increased dark state current after illumination in applications such as a light activated memory ("light memory device").

The spectral response of the BCB-photOFET was estimated. Amplification of the photogenerated currents in the strong accumulation regime except for a suppression of photocurrent induced by photons in the blue spectral region, coinciding with a threshold voltage shift as already observed under white-light illumination. This leads to the conclusion that the absorption of UV-photons induces reversible changes in the device transfer characteristics than can be utilized in a light memory device.

5.6 References

[1] N. Marjanović, Th. B. Singh, G. Dennler, S. Günes, H. Neugebauer, N. S. Sariciftci, R. Schwödiauer, and S. Bauer, *Org. Elect.* in press

[2] K. S. Narayan and N. Kumar, Appl. Phys. Lett. 79, 1891, 2001.

[3] M. C. Hamilton, S. Martin, and J. Kanicki, *IEEE Trans. Electron Devices*, 51 877, 2004.

[4] T. P. I Saragi, R. Pudzich, T. Fuhrmann, and J. Salbeck, *Appl. Phys. Lett.* 84, 2334, 2004.

[5] Y. Xu, P. R. Berger, J. N. Wilson, and U. H. F. Bunz, *Appl. Phys. Lett.* 85, 4219, 2004.

[6] Th. B. Singh, N. Marjanović, P. Stadler, M. Auinger, G. J. Matt, S. Günes, N. S. Sariciftci, R. Schwödiauer, and S. Bauer, *J. Appl. Phys.* 97, 083714, 2005.

[7] Th. B. Singh, N. Marjanović, G. J. Matt, N. S. Sariciftci, R. Schwödiauer, and S. Bauer, *Appl. Phys. Lett.* 85, 5409.

[8] L.-L. Chua, J. Zaumsell, J.-F. Chang, E. C.-W. Ou, P. K.-H. Ho, H. Sirringhaus, and R. H. Friend, *Nature* 434, 194, 2005.

[9] A. Sallelo, M. L. Chabinyc, M. S. Yang, and R. A. Street, *Appl. Phys. Lett.* 81, 4383, 2002.

[10] A. Selleo and R.A. Street, *J. Appl. Phys.* 94, 471, 2003.

[11] S. Dutta and K.S. Narayan, *Appl. Phys. Lett.* 87, 193505, 2005.

[12] M. Breban, D.B. Romero, S. Mezhenny, V. W. Ballaratto, and E.D. Williams, *Appl. Phys. Lett.* 87, 203503, 2005.

[13] J.-M. Choi, J. Lee, D.K. Hwang, J.H. Kim, S. Im, and E. Kim, *Appl. Phys. Lett.* 88, 043508, 2006.

Chapter 6

6. Memory Elements based on Organic Field-Effect Transistors and Light Memory Element based on Photoresponsive Organic Field-Effect Transistors

In this chapter, a non-volatile memory device based on an OFET using a polymeric electret as gate dielectric and light memory device based on the trapping of photoinduced charges in BCB-photOFETs are presented.

6.1 Non-volatile Organic Field-Effect Transistor Memory Element with a Polymeric Gate Electret

A typical OFET structure is shown in Fig. 6.1 together with the chemical structure of PVA and PCBM [1]. The details about the device fabrication are given in Chapter 2. The channel length, L of the OFET is 65 μm and the channel width $W = 1.4$ mm. For the electret, a thickness $d = 1.4$ μm, dielectric constant, $\varepsilon_{pva} = 5$ and capacitance of $C_{PVA} = 3$ nF/cm^2 was measured. This results in a d/L ratio ≈ 0.02, which is acceptable in order to avoid having the gate field screened by the source-drain contacts. For both source and drain electrodes, Cr is used.

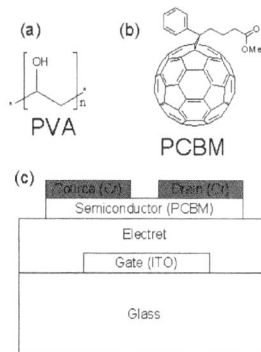

Fig. 6.1 Chemical structure of (a) PVA, (b) methanofullerene (PCBM), and (c) schematic of the staggered mode nonvolatile memory OFET.

The transistor characteristics $I_{ds}(V_{ds})$ are shown in Fig. 6.2 featuring an n-channel FET [2-4] with electron accumulation mode with applied positive V_{gs} and an electron depletion mode with increasing negative V_{gs} [1]. A saturation of I_{ds} with increasing V_{ds} is obtained even if no V_{gs} is applied. It is not clear as yet why FETs based on fullerenes have this "on" behaviour with $V_{gs} = 0$ V. The substrate surface may play a critical role in this effect; devices with substrates treated with amines have also been found to show this behaviour (Fig. 6.2) [5].

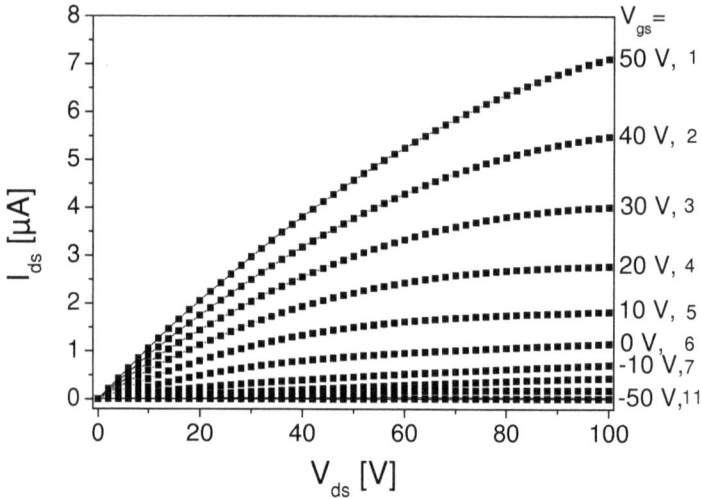

Fig. 6.2 Transistor characteristics of an OFET with channel length $L = 65$ μm, channel width $W = 1.4$ mm for different V_{gs}. All measurements were carried out at room temperature. The data shown here are taken in descending V_g mode from 50 to –50 V in steps of 10 V as labelled from 1- 11. The integration time is 1 s.

We observed a difference in $I_{ds}(V_{ds})$ plots while taking the data by increasing and/or descending the gate voltages. Data presented here are recorded when the gate voltage is applied in descending mode with 1s integration time. The sequence of the measurement in Fig. 6.2 is labelled from 1-11. Transistor characteristics of the devices show amplification factors up to 10^4 with gate on/off.

The large hysteresis observed in V_g is shown in the transfer characteristics (I_{ds} *versus* V_{gs}) cycling the gate voltage (Fig. 6.3) [1]. All the OFET devices reported herein show a sharp threshold voltage V_{th}, a voltage which is the x-intercept of the plot of $\sqrt{I_{ds}}$ vs V_{gs}. Initial cycles feature a negative V_{th}, which develops into a stable hysteresis after few cycles, with turn-on around 0 V. This allows us to calculate the mobility, μ from the $\sqrt{I_{ds}}$ vs. V_{gs} plot as shown in the inset of Fig. 6.3. A value of $\mu = 9 \times 10^{-2}$ cm^2/Vs is obtained using Equ. (2) (see Chapter 1).

A μ of approximately 10^{-1} cm^2/Vs is relatively high as compared to the reports on PCBM devices using space charge limited currents [6, 7] and field effect studies [8]. The origin of this improved mobility here is proposed to be the homogenous film formation on top of the smooth electret PVA. We did not observe any significant dependence of this mobility upon variation of the source–drain metal electrodes like calcium and LiF/Al.

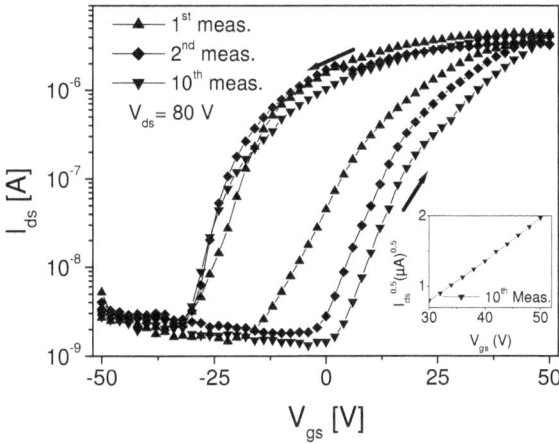

Fig. 6.3 Transfer characteristics of the OFET with V_{ds}=80 V demonstrating the nonvolatile organic memory device. Each measurement was carried out with an integration time of 1 s. Inset: $I_{ds}^{0.5}$ vs. V_{gs} plot for the 10th measurement.

As presented in Fig. 6.3 the magnitude of the source-drain current I_{ds} increases with an amplification of up to 10^4 at $V_{gs} \approx 50$ V with respect to the initial "off" state with $V_{gs} = 0$ V. However, the saturated I_{ds} remain at high values even when V_{gs} reduces back to $V_{gs} = 0$ V. In order to completely deplete I_{ds}, one needs to apply a reverse voltage of $V_{gs} \approx -30$ V. A large shift in V_{th} by 14 V is observed when measured for the second time with respect to the initial cycle. Compared to the 2nd measurement, the 10th measurement has not shown a significant shift in V_{th}. Each measurement was performed with a long integration time of 1 s with a step voltage of 2 V. Our result indicates that there is minimal gate bias stress in these devices [9]. This observation is proposed to be due to locally trapped charges, which induce shifts in V_{th}.

To estimate the retention time of the stored charges remaining in the electret (*i.e.* storage time of the memory element), time-resolved measurements were performed (Fig.

6.4(a) and 6.4(b)) [1]. First the device was biased with V_{ds}= 80 V and kept at floating gate. At time t = 0 s, V_{gs} = 50 V is applied until a stable current is obtained. After a time t = 500 s the device is left with a floating gate. I_{ds} remain high (memory "on" state) for more than 15 h. This implies that once the electret is charged fully, the relaxation of the charges is a slow process as expected for charged electrets [10]. No detectable degradation in the devices properties has been observed after this long measurement time. In Fig. 6.4(b) the write/read/erase/read cycles are demonstrated with write/erase pulses using positive and negative gate voltages, respectively. Monitoring the I_{ds} correspond to reading the memory state. High I_{ds} denotes the "on" and low I_{ds} the "off" state of the memory unit. The device presents a quite long response time due to the electret mechanism of charge storage, as discussed subsequently, but improvements seem possible with thinner gate dielectrics.

Our results cannot be explained by a dipole polarization mechanism of the electret [11], since capacitance voltage measurement showed a negligible hysteresis. Therefore trapping of injected charges is proposed.

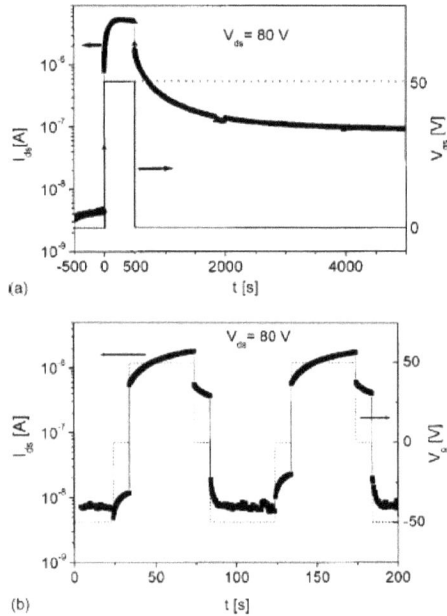

Fig. 6.4 (a) Logarithmic I_{ds} vs. time (t) plot at V_{ds}= 80 V for floating gate (denoted by V_{gs} = 0 V), during V_{gs} = 50 V and floating gate sequentially showing long retention time. The data were taken each 250 ms and every second data point is plotted. (b) Switching response of the drain current, upon application of gate voltage pulses with a pulse height of ± 50 V and a pulse duration of 40 s.

6.2. Light Memory Element based on the BCB/MDMO-PPV: PCBM (1:4) blends based photOFETs

After illumination of the BCB-photOFET with a white light source or blue/UV light (see Chapter 5, § 5.3.3), a significant shift of the dark transfer curve with respect to the initial dark transfer curves is observed, presumably due to several effects which may occur at the BCB/MDMO-PPV: PCBM blend interface. Recovery of the initial dark state in that system was achieved by annealing. Narayan et al reported earlier similar effects in pristine poly (3-hexylthiophene) (P3HT) based OFETs fabricated on PVA [12].

The effect of an increased dark state current after illumination was studied in BCB-photOFETs. Details about the device fabrication and characterization were already given earlier (Chapter 2). In Figure 6.5, the switching response of the drain-source current of a BCB-photOFET upon application of gate voltage and light pulses is shown. The switching cycles are achieved using white light illumination (AM1.5, 100 mW.cm^{-2}) and read/erase pulses using positive (+20 V) and negative gate voltages (-50 V), respectively. Monitoring I_{ds} reads the memory state. By biasing the device with +20 V, the transistor is in the ON-state, which corresponds to the reading state "0". Upon illumination, I_{ds} increases due to photodoping, this corresponds to "writing". After switching off the light charge trapping effects at the BCB/blend interface causes persistent photodoping, which corresponds to the memory state "1". Biasing the device with a high negative gate voltage, -50 V, sets the memory state back to "0". Fig. 6.6 shows the switching response of the light memory element based on the BCB-photOFETs for the first three cycles.

The "light induced memory" experiment was also performed under monochromatic illumination with $\lambda = 514$ nm (300 mW.cm^{-2}), Fig. 6.7. Absence of the light induced memory effect at that illumination conditions was observed, indicating that the presence of blue/UV irradiation is necessary.

It is worth to mention here that light induced memory effect was not observed in pristine PCBM or MDMO-PPV based photOFETs.

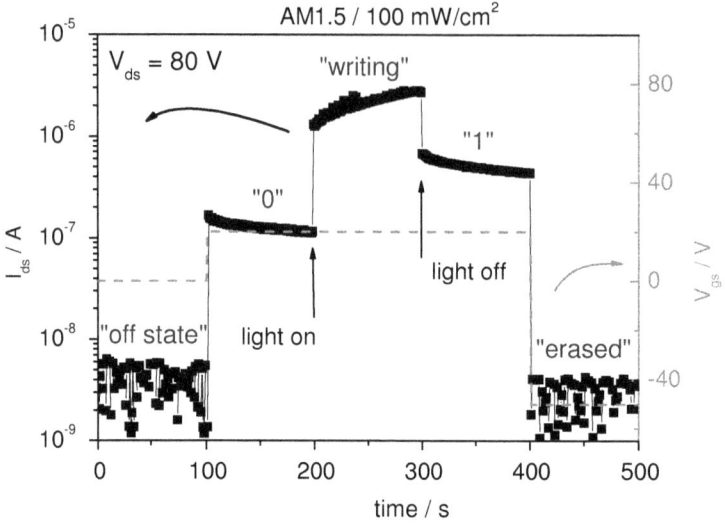

Fig. 6.5 Switching response of the drain-source current of a BCB-photOFET upon application of gate voltage pulses with pulse heights of + 20 V and –50 V and pulses of white light (AM1.5 / 100 mW.cm^{-2}) with a duration of 100 seconds.

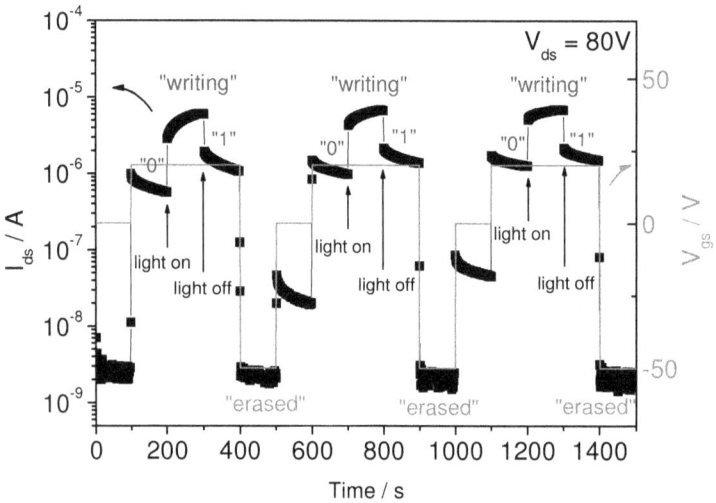

Fig. 6.6 Switching response of the light memory element based on BCB-photOFETs, under AM1.5 / 100 mW.cm^{-2} illumination, for the first three cycles.

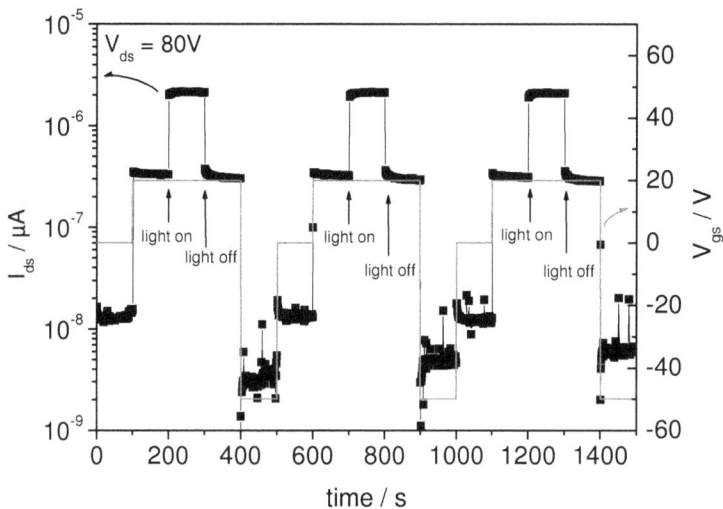

Fig. 6.7 Switching response of the drain-source current of a BCB-photOFET upon application of gate voltage pulses with pulse heights of + 20 V and –50 V and pulses of monochromatic light (λ = 514 nm / 300 mW.cm^{-2}) with a duration of 100 seconds, for the first three cycles.

As known, the threshold voltage shift is commonly attributed to a built-in electric field near the dielectric/semiconductor interface caused by the presence of a sheet of charges [10, 13]. The trapped charges screen the gate field, which in turn causes the threshold voltage shift. Removing (detrapping) the trapped charge leads to the recovery of the system to their initial state. Based on the above mentioned and results presented in Chapters 5, it is assumed that in the BCB/MDMO-PPV: PCBM (1:4) blend based systems, the long living trapped charges originating from the excitation of the conjugated polymer/fullerene blend in the blue region of the visible spectrum are responsible for the "light induced memory" effect. For a full understanding of the observed effects, further investigations have to be taken.

6.3. Summary

In summary, we demonstrated an organic non-volatile memory device based on OFETs using a polymeric electret as gate dielectric. The results indicate metastable charging of the electret with an applied gate voltage resulting in very long retention times up to hours.

A light memory element was realized in BCB-photOFETs upon illumination with white light pulses and simultaneously biasing the transistor with gate voltage pulses. "Writing" was done by illumination of the device and "reading/erasing" was done by applying positive or large negative gate voltage pulses. No light memory effect was observed in the BCB-photOFET illuminated with monochromatic light source ($\lambda = 514$ nm).

It was observed that photoexcitation of the blend in the blue region is responsible for the long living charge traps which causes the "light induced memory" effect. Further experiments are proposed for fully understanding the observed phenomenon.

6.4. References

[1] Th. B. Singh, N. Marjanović, G. J. Matt, N. S. Sariciftci, R. Schwödiauer, and S. Bauer, *Appl. Phys. Lett.* 85, 5409, 2004.

[2] T. Kanbara, K. Shibata, S. Fujiki, Y. Kubozono, S. Kashino, Y. Kubozono, S. Kashino, T. Urishi, M. Sakai, A. Fujiwwara, R. Kumashiro, and K. Tanigaki, *Chem. Phys. Lett.* 379, 223 2003.

[3] S. Kaboyashi, S. Mori, S. Lida, H. Ando, T. Takenobu, Y. Taguchi, A. Fujiwara, A. Taninaka, H. Shinohara, and Y. Iwasa, *J. Am. Chem. Soc.* 125, 8116, 2003.

[4] K. Shibata, Y. Kubozono, T. Kanabara, T. Hosokawa, T. Hosokawa, A. Fuziwara, Y. Ito, and H. Shinohara, *Appl. Phys. Lett.* 84, 2572, 2004.

[5] R. C. Hoddon, A. S. Perel, R. C. Morris, T. T. M. Palstra, A. F. Hebard, and R. M. Fleming, *Appl. Phys. Lett.* 67, 121, 1995.

[6] V. D. Mihailetchi, J. K. J. van Duren, P. W. M. Blom, J. C. Hummelen, R. A. J. Janssen, J. M. Kroon, M. T. Rispens, W. J. H. Verhees, and M. M. Wienk, *Adv. Func. Mater.* 13, 43, 2003.

[7] G. J. Matt, N. S. Sariciftci, and T. Fromherz, *Appl. Phys. Lett.* 84, 1570, 2004.

[8] C. Waldauf, P. Schilinsky, M. Perisutti, J. Hauch, and C. J. Brabec, *Adv. Mater.* 15, 2084, 2003.

[9] A. Sallelo, M. L. Chabinyc, M. S. Yang, and R. A Street, *Appl. Phys. Lett.* 81, 4383, 2002.

[10] R. Gerhard-Multhaupt and G. Sessler, *Electrets*, Vol. I and II, (Laplacian Press, Morgan Hill, 1999).

[11] H. E. Katz, X. M. Hong, A. Dodabalapur and R. Sarpeshkar, *J. Appl. Phys.* 91, 1572, 2002.

[12] S. Dutta and K.S. Narayan, *Adv. Mater*, 16, 2151, 2004.

[13] S.M Sze, *Physic of Semiconductor Devices*, Wiley-Interscience, New York, 1981.

Chapter 7

7. Summary and Outlook

In this work, a systematic study on photOFETs based on conjugated polymer/fullerene blends as the photoactive semiconductor layer and poly-vinyl-alchocol (PVA) or divinyltetramethyldisiloxane-bis(benzocyclobutene) (BCB) as highly transparent polymeric gate dielectrics is reported.

PhotOFETs fabricated on PVA show high responsivity but weak photostability, whereas photOFETs fabricated on BCB show transistor behaviour in a broad range of illumination intensities and good photostability even at high illumination intensities. For devices with both dielectrics, the observed increase in the drain-source current under illumination suggests the generation of carriers in the bulk of the highly photoactive conjugated polymer/fullerene blend.

PhotOFETs fabricated on top of cross-linked BCB as dielectric show phototransistor behaviour with sufficiently good photostability. A shift of the dark transfer curve with respect to the initial dark transfer curves was observed, presumably due to charge trapping which might occur at the BCB/blend interface. Recovery of the initial dark state in this devices was achieved either by annealing or by applying a high negative gate voltage. From the transfer characteristics in the dark taken sequentially during the optoelectrical measurement, it was concluded that the photoexcitation in the blue (high energy photons) is mainly causing charge trapping, which results in an increased dark state current after illumination ("light induced memory").

The photoresponsivity of organic field effect transistors (photOFETs) is interesting since it is the basis for light sensitive transistors. PhotOFETs can be used *e.g.* for light induced switches, light triggered amplification, detection circuits and, in photOFET arrays, for highly sensitive image sensors. PhotOFETs based on conjugated polymer/fullerene mixture and organic dielectrics, together with a non-volatile memory element, presented in this work have potential to be used in such applications.

86